Hepatitis B

The Virus, the Disease, and the Vaccine

Hepatitis B

The Virus, the Disease, and the Vaccine

Edited by

Irving Millman

Institute for Cancer Research
Fox Chase Cancer Center
Philadelphia, Pennsylvania

Toby K. Eisenstein

Temple University School of Medicine
Philadelphia, Pennsylvania

and

Baruch S. Blumberg

Institute for Cancer Research
Fox Chase Cancer Center
Philadelphia, Pennsylvania

Springer Science+Business Media, LLC

Library of Congress Cataloging in Publication Data

Main entry under title:

Hepatitis B: the virus, the disease, and the vaccine.

"Proceedings of a symposium . . . held November 11-12, 1982, in Philadelphia, Pennsylvania"—Verso t.p.

Sponsored by Eastern Pennsylvania Branch, American Society for Microbiology and others.

Includes bibliographical references and index.

1. Hepatitis B vaccine—Congresses. 2. Hepatitis B—Congresses. 3. Hepatitis viruses—Congresses. I. Millman, Irving, 1923- .II. Eisenstein, Toby K. III. Blumberg, Baruch S. IV. American Society for Microbiology. Eastern Pennsylvania Branch.

QR189.5.H46H47 1984 616.3′623 84-11510

DOI 10.1007/978-1-4899-0369-3

Proceedings of a Symposium on Hepatitis B: The Virus, the Disease, and the Vaccine, held November 11-12, 1982, in Philadelphia, Pennsylvania

Originally published by Plenum Press, New York in 1984
MyCopy version of the original edition 1984

SPONSORS OF THE THIRTEENTH ANNUAL

CLINICAL MICROBIOLOGY SYMPOSIUM

Eastern Pennsylvania Branch, American Society for Microbiology

Merck Sharp & Dohme

Bureau of Laboratories, Pennsylvania Department of Health

Department of Microbiology and Immunology, Temple University School of Medicine

Department of Microbiology and Immunology, Hahnemann University

Department of Microbiology, Thomas Jefferson University

Department of Microbiology, University of Pennsylvania School of Medicine

Department of Microbiology, Medical College of Pennsylvania

ORGANIZING COMMITTEE

Conference Co-chairmen

Baruch S. Blumberg
Institute for Cancer Research
Fox Chase Cancer Center
Philadelphia, PA

Irving Millman
Institute for Cancer Research
Fox Chase Cancer Center
Philadelphia, PA

Symposium Committee Chairperson

Toby K. Eisenstein
Temple University School of Medicine
Philadelphia, PA

Committee

Carl Abramson
Pennsylvania College of
Podiatric Medicine
Philadelphia, PA

Josephine Bartola
Bureau of Laboratories
Pennsylvania Department of Health
Lionville, PA

Kenneth R. Cundy
Temple University
School of Medicine
Philadelphia, PA

Morton Klein
Temple University
School of Medicine
Philadelphia, PA

Linda Miller
Bryn Mawr Hospital
Bryn Mawr, PA

Norman P. Willett
Temple University
School of Dentistry
Philadelphia, PA

ACKNOWLEDGMENTS

The editors are indebted to the Eastern Pennsylvania Branch of the American Society for Microbiology under whose sponsorship this conference was held, and to the members of the organizing committee whose hard work made the symposium possible.

We would also like to thank Dr. Henry Bielstein, the branch President for his encouragement and enthusiastic support of this endeavor. The Pennsylvania Department of Health-Bureau of Laboratories generously handled mailings and registrations.

We are also grateful to Temple University School of Medicine, Hahnemann Medical College, Thomas Jefferson Medical College, and the School of Medicine of the University of Pennsylvania for their sponsorship. The Department of Microbiology and Immunology of Temple University and the Division of Clinical Research of the Institute for Cancer Research were particularly supportive of this conference and this volume by providing office services.

This conference would not have been possible without the generous financial support of Merck Sharp and Dohme whose contribution we gratefully acknowledge.

The excellent proofreading skills of Joyce Codispoti and Peggy Nowak are appreciated. Finally, many thanks to Maureen Walsh for her beautiful camera-ready copy of the manuscripts.

CONTENTS

INTRODUCTION

Toby K. Eisenstein
Symposium Committee Chairperson
Temple University School of Medicine
Philadelphia, Pennsylvania 19140

This symposium is the thirteenth biennial clinical microbiology program sponsored by the Eastern Pennsylvania Branch of the American Society for Microbiology in cooperation with the Philadelphia area medical schools and the Bureau of Laboratories of the Pennsylvania Department of Health. This year a generous contribution from Merck, Sharp and Dohme has helped to make the program a reality.

The subject matter for this symposium represents an attractive spectrum of medical, biological and molecular approaches to the practical solution of a public health problem--namely, prevention of infection with the hepatitis B virus. The symposium may be unique in that it focuses on a product which was first marketed less than three months ago, but included in the program are presentations on two new approaches to hepatitis B vaccine production which may replace the one which is newly unveiled. The rapidity of progress in our present era of biological research is indeed astonishing.

Vaccine development has been the major application, for human benefit, of research in microbiology and immunology. From Jenner's empirical observations in 1796 on protection of milkmaids against smallpox by cowpox infection, we have witnessed, through vaccination, the extinction in the 20th century of this ancient scourge of mankind (1). There are presently many novel and exciting approaches for improving existing vaccines, and for preparing new ones for organisms against which prophylaxis was never before available. The hepatitis B story illustrates how ingenuity in approach brought a solution to the problem of how to obtain an antigen which cannot be grown in tissue culture or laboratory animals.

In the case of the pneumococcus, Group B streptococcus, and Hemophilus influenzae, highly purified capsular polysaccharides are being used or developed as nontoxic vaccines (2,3,4). Biochemical coupling of antigenic determinants to carrier proteins, such as meningococcal polysaccharide to tetanus toxoid (5) and detoxified lipopolysaccharide of Pseudomonas aeruginosa to toxin A (6) are examples of molecular engineering applied to vaccine research. The isolation of the Texas Star strain of Vibrio cholerae, a mutant which produces the B or binding subunits of cholera toxin but not the A or toxic moiety, is an example of exquisite selection techniques to find the proper strain based on our appreciation of the molecular mechanisms of disease causation (7). For veterinary use, a chemically synthesized peptide vaccine has been produced against foot and mouth disease by sequencing the genome of the virus (8). In the last session of this symposium, we will hear about a synthetic hepatitis antigen, and also production of the hepatitis core antigen by cloning the gene in E. coli. Thus, vaccine development is currently an area of intensive investigation, where the newest methods are being applied to solve some of mankind's oldest problems.

For this symposium, we have gathered together researchers, clinicians and epidemiolgists to describe how each of these disciplines has contributed to the production of a vaccine against hepatitis B in the comparatively short span since discovery that the virus is the etiologic agent of the disease.

I know you will find the proceedings informative and exciting, so on behalf of the Eastern Pennsylvania Branch, I welcome you all.

REFERENCES

1. Henderson, D. A. (1976). The eradication of smallpox. Sci. Amer. 235:25.
2. Austrian, R. (1981). Pneumococcus: The first one hundred years. Rev. Infect. Dis. Suppl. 3:183.
3. Carey, R. B., Eisenstein, T. K., Shockman, G. D., Greber, T. F., and Swenson, R. M. (1980). Soluble group- and type-specific antigens from type III group B Streptococcus. Infect. Immun. 28:195-203.
4. Robbins, J. B., Schneerson, R., and Parke, J. C., Jr. (1982). A review of the efficacy trials with Haemophilus influenzae type b polysaccharide vaccines, in Haemphilus influenzae, S. H. Sell and P. F. Wright, eds., Elsevier, New York, pp. 255-264.
5. Jennings, H. J. and Lugowski, C. (1982). Tetanus toxoid conjugates of the meningococcal polysaccharides, in Seminars in Infectious Disease, Vol. IV: Bacterial Vaccines, J. B. Robbins, J. C. Hill and J. C. Sadoff, eds., Thieme-Stratton, New York, pp. 247-253.

6. Sadoff, J. C., Futrovsky, S. L., Sidberry, H. F., Iglewski, B. H., and Seid, R. C., Jr. (1982). Detoxified lipopolysaccharide-protein conjugates, in Seminars in Infectious Disease, Vol. IV: Bacterial Vaccines, J. B. Robbins, J. C. Hill and J. C. Sadoff, eds., Thieme-Stratton, New York, pp. 346-354.
7. Honda, T. and Finkelstein, R. A. (1979). Selection and characteristics of a Vibrio cholerae mutant lacking the A (ADP-ribosylating) portion of the cholera enterotoxin. Proc. Natl. Acad. Sci. USA 76:2052.
8. Bittle, J. L., Houghten, R. A., Alexander, H., Shinnick, T. M., Sutcliffe, J. G., and Lerner, R. A. (1982). Protection against foot-and-mouth disease by immunization with a chemically synthesized peptide predicted from the viral nucleotide sequence. Nature 298:30-33.

KEYNOTE ADDRESS: THE AUSTRALIA ANTIGEN STORY

Baruch S. Blumberg

Institute for Cancer Research
Fox Chase Cancer Center
Philadelphia, Pennsylvania 19111

I would like to welcome visitors to Philadelphia and remind them that we are in the midst of celebrating our 300th birthday and looking forward to our fourth century as a thriving city. For the past year the city has undertaken an orgy of reminiscences and the commemoration of historic events. Philadelphia is gifted in preserving and recalling its past, and we would like to think that we are also interested in the development of an exciting future. All this historical reminiscing may provide an adequate excuse to look back at the work that we have done at the Institute for Cancer Research over the past 18 years which led to the discovery of the hepatitis B virus, the invention of the vaccine to protect against it, the possibility of prevention of primary cancer of the liver, and the many developments in our knowledge of this interesting virus. I hope that our extended discussion of the past won't detract from an interest in our current work, to which I will also refer briefly.

This work was accomplished by many investigators in our Institute. Figure 1 shows some of these. It is a photograph taken during September 1980 shortly after a site visit for one of our NIH program project grants. We have been fortunate in having an intelligent, dedicated and congenial group of scientists and staff working in our laboratory. It has been a great pleasure to be associated with them.

In this paper I plan to review our investigations beginning with the finding of Australia antigen and its identification as the surface antigen of hepatitis B virus. The narrative will proceed approximately chronologically, but themes will be

Figure 1. The staff of the Division of Clinical Research, Institute for Cancer Research, September 19, 1980.

developed out of linear time and into their eventual outcome. There will also be a digression to examine in detail how a scientific discovery, the identification of HBV carriers, became accepted into general medical and public health practice.

DISCOVERY OF AUSTRALIA ANTIGEN

In 1963 a major interest in our laboratory was the study of human biochemical and immunologic variation. A fundamental question that faces the physician is that of why some people become ill and others remain healthy even though all are exposed to the same disease hazard. Clearly, some of this is a consequence of chemical and immunologic variation in humans. We started in 1956 to study variation in serum proteins using the newly introduced starch gel electrophoresis method. We soon learned from studies in British, Basque, African, Alaskan and other populations that there was indeed a considerable polymorphic variation in several serum proteins (see, for example, references 1 and 2). We then made the hypothesis that if some of these serum protein variants were antigenic, transfused patients might develop detectable antibodies in their serum against variants which they had not inherited or acquired. We employed the method of double diffusion in agar gel using sera from transfused patients as the source of the putative antibody and testing these against other sera from normal people. Using this technique, we found a precipitating antibody that identified a complex inherited system of serum low density lipoproteins which have since become of interest in genetic, anthropologic, forensic and other fields (3). The hypothesis of antigenic polymorphism had been supported by this observation, and we continued to test the hypothesis further by using sera from additional transfused patients to test against sera obtained from other populations. Since we were looking for unknown polymorphisms and the allele frequencies for most human polymorphisms vary greatly from population to population, we included in the serum panel against which the transfused sera were to be tested not only local populations but also those from Africa, Asia, Australia and elsewhere.

During the course of this ongoing research, a precipitin reaction dissimilar from any seen before was observed; and this reaction was between the serum from an Australian aborigine and that of a frequently transfused hemophilia patient from New York City (4). Figure 2 is an illustration, taken from an early publication illustrating such a precipitin reaction. (This is not the original Australia aborigine/hemophilia band, for which we do not apparently have a photograph.) What was this new phenomenon? What was the character and significance of "Australia antigen" (abbreviated Au), as we termed the protein present in the aborigine? In order to find out, it was necessary to formulate

Figure 2. Precipitin reaction between serum from patient with "Australia antigen" (HBsAg) (top) and serum from hemophilia patient containing antibody against the antigen (anti-HBs) (bottom). Adapted from the first paper illustrating this reaction (4). (Reprinted by permission from JAMA 191:541-546, copyright 1965.)

a series of testable hypotheses, and additional observations were required to do this. Australia antigen was stable in sera kept in a frozen state, and we were able to test thousands of these taken from the large collection housed at the Division of Clinical Research of the Institute for Cancer Research. We learned that Au was very rare in U.S. populations but common (about 5-10%) in some African, Asian and Oceanic groups. We also learned that it was common in leukemia patients, most of whom had been transfused. Based on this observation, we made a series of hypotheses including the hypothesis that there is an inherited trait which makes people susceptible both to leukemia and to persistent carriage of the Australia antigen. To test this we generated a corollary hypothesis; namely, that people who have a high risk of developing leukemia should also have a high frequency of Au. Several such groups are known. Children with Down's syndrome (DS, mental retardation associated with trisomy 21) have a 20 fold or greater risk of developing leukemia. We tested groups of institutionalized DS patients and compared them to other mentally retarded children in the same institution (5). In all cases the frequency of Au was high in the DS patients ($\sim$ 30%) and much lower in the controls. This result was gratifying in that it not only fulfilled the predictions from the hypothesis, but also allowed us to observe a group of individuals who were closer to home than the Australian aborigines and other populations in whom a high frequency of Au was found. We learned that the presence or absence of Au appeared to be a persistent trait; if Au was present at first testing, then it was likely to be present at all subsequent tests. Similarly, if Au was absent on first testing, then it was absent on subsequent testing. Early in 1966 we found an exception to the latter. One of our DS patients (J.B.), who previously had not had Au, did have it on a subsequent testing. He was admitted to our Clinical Research Unit at the Jeanes Hospital (which adjoins the Institute for Cancer Research) for observation and evaluation. What appeared to be a "new" protein had appeared in his serum, and since most serum proteins are manufactured in the liver, we performed a series of "liver function" tests. These showed that between the testing in which Au was not found in his blood and the "positive" test, he had developed a form of chronic anicteric hepatitis which could be detected only by measurement of the changes in his blood.

This observation, then, generated the hypothesis that Au was associated with "viral hepatitis." For many years this clinical syndrome was assumed to be of viral etiology, but the virus itself had not been identified. We tested the association hypothesis by requesting specimens of blood from patients with the clinical diagnosis of hepatitis for our clinical colleagues. We soon were able to establish that there was a much higher frequency of Au in the hepatitis patients, both acute and chronic, than in the controls. We also completed a systematic analysis of the Down's syndrome patients with and without Au and found that the former had significantly higher levels of SGPT and other evidences of liver abnormality than the latter (6). With this support for the association hypothesis, we then proceeded to test the concept that Au was, or was on, the hepatitis virus. With the collaboration of our colleague Manfred Bayer of our Institute, we visualized particles with the appearance of a virus in a series of electron microscope studies (Figures 3a and b); and a variety of other studies indicated that these particles, which reacted with the antibody against Au, were hepatitis viruses (7). We found that these particles did not have nucleic acid and it later eventuated that they were a part of the virus which consisted only of the surface antigen.

Early in 1969 it became apparent to us that a vaccine could be produced from the peripheral blood of carriers. We postulated the existence of a whole virus particle (and this was later seen and identified by Dane and his colleagues (8)) and proposed that the particles containing only the surface antigen of the hepatitis virus (the same small particles we had originally visualized in the EM) could be used as an immunogen. Our epidemiologic observations indicated that people with anti-HBs (as the antibody against the surface was later designated) did not usually become infected with HBV. More convincing evidence came from the studies of Professor Kazuo Okochi who was then in Tokyo (9). He found that patients who had anti-HBs when transfused with Au (i.e. hepatitis virus) were much less likely to develop hepatitis than those who did not have

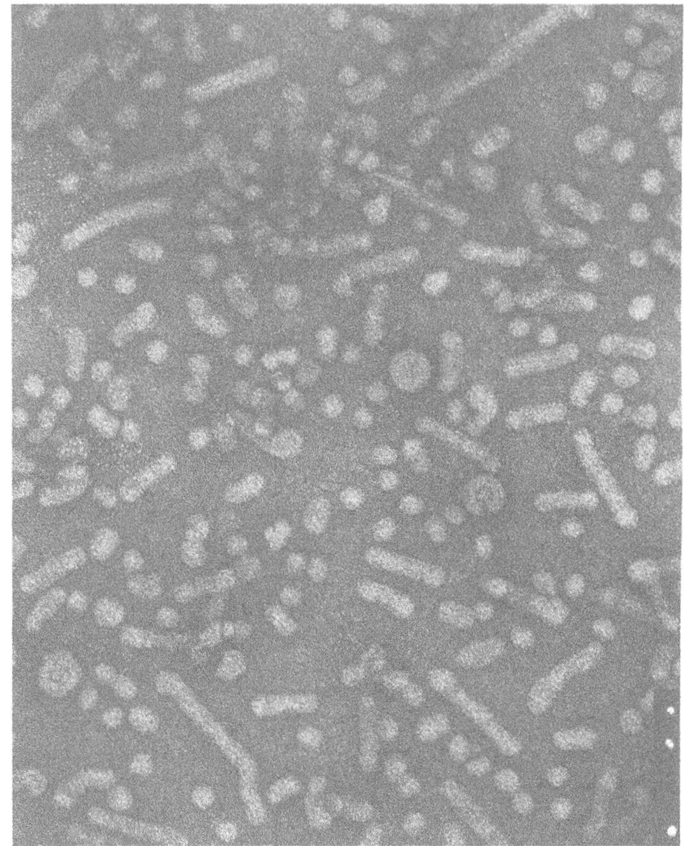

Figure 3a. Electron micrograph of hepatitis B virus showing whole viruses (the large spheres), surface antigen particles (the small spheres), and elongated surface antigen particles (magnification approximately 300,000 x).

antibody; that is, anti-HBs was protective. Dr. Irving Millman, who had joined our group in 1967, had, by what in retrospect was a remarkable coincidence, a considerable amount of experience in the invention of vaccines. We devised a method for separating the surface antigen particles from the postulated whole virus using centrifugation and other physical-chemical means. A United States patent for this product was applied for in 1969 and issued in 1971 (10). In that year we began negotiations with Merck & Company whose vaccine development facilities were located near Philadelphia. After extensive trials, including those of Szmuness, Hilleman and others, the vaccine was approved by the Food and Drug Administration

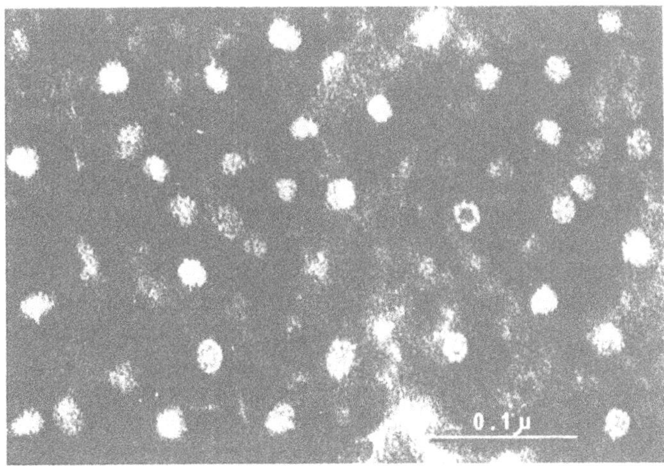

Figure 3b. Electron micrograph from the original publication (7) showing only the surface antigen particles from which the vaccine is prepared. Electron micrographs courtesy of M. Bayer. (Reprinted by permission from NATURE 218: 1057-1059, copyright 1968 (MacMillan Journals Limited.))

in 1981 and came into general use by the end of that year. Irving Millman has prepared a detailed account of the invention and development of the hepatitis vaccine, which is presented later in this symposium.

A series of investigators (Okochi, Prince, Vierucci and others) soon confirmed the identification of Au with the hepatitis virus and established that it was specifically associated with the hepatitis B virus. It became apparent that the immunodiffusion test for Au could be used to diagnose the presence of hepatitis B virus (HBV) in serum and other biological fluids. More importantly, from the viewpoint of prevention, it could be used to detect occult hepatitis carriers among blood donors.

APPLICATION OF BASIC RESEARCH TO CLINICAL PRACTICE

An important question that often arises in contemporary research is how basic investigation became applied. It may be useful to examine in detail the steps which were taken between the time the original observations on Australia antigen were made and the final use of these findings as a medical and public health measure for the detection of blood donor carriers. Earlier in the text we described the discovery of Australia antigen, the change in Au reactivity in the Down's syndrome patient coincident with the development of chemical evidence of hepatitis, and the tests of the hypothesis that Au is associated with hepatitis, which was supported. Later we tested the hypothesis that Au is on the hepatitis virus, and this was also supported. During this period (late in 1966), we had been considering the role of Australia antigen in donor blood testing and in November 1966 I recorded in my notebook a discussion of hepatitis and transfusion. By October 1968 the ICR laboratory was systematically distributing reagents for the "Australia antigen test" to laboratories throughout the world. In January 1969 the clinical protocol we had devised to test the hypothesis that Au in a donor's blood increases the probability of post-transfusion hepatitis in the blood of its recipients was approved by our Clinical Investigation Review Committee, and the study began immediately afterward. However, in June 1969 the clinical data collected by Kazuo Okochi in Tokyo became available to us (11). Okochi came to our laboratory after an international meeting convened at Yale University to discuss the recent findings on Au and, in particular, on its nomenclature. In a study similar to the one in progress in our laboratory, he had compared the development of post-transfusion hepatitis in patients who received donor blood containing Au to patients who received donor blood which did not contain Au. He found that there was a significantly higher frequency of post-transfusion hepatitis in the former than the latter. Based on the Okochi data and our confidence in his methods and results, we decided that the hypothesis had been amply supported (given the impressive evidence that preceded it that Au was on the hepatitis virus); and we decided to abort our study.

Although we had aborted the original research study, we continued to test the blood samples sent to us by the blood bank of the Philadelphia General Hospital and informed them as soon as an Au positive donor was detected in order for them to forestall the use of this blood. It was at this point that the exercise stopped being a research project and became a clinical application. Although our original research study was stopped, we were later able to compare the post-transfusion hepatitis experience at Philadelphia General Hospital before and after the donor testing program was instituted. Hepatitis was decreased by two thirds of its original level, presumably as a result of the testing procedure (12).

Although we were convinced that this technique could be applied, it was some time (in fact, a relatively short time) before it gained general acceptance. What happened between the time we were convinced and others became convinced of its clinical application? A detailed investigation of what occurred may be useful in understanding how research results achieve general acceptance. The facts of which we are aware that appeared to have a bearing on this process will now be reviewed.

Curiously, there does not appear to have been any large amount of additional scientific data which changed convictions; rather, a series of meetings, administrative steps and certain legal procedures occurred during this interval. In June 1969 we had published a short note in the Bulletin of Pathology inferring that testing might be appropriate (13). In May 1969 Gocke and Kavey published the results of a preliminary transfusion study involving eight patients which also supported the hypothesis (14). On October 31 and November 1, 1969, a panel of the Committee on Plasma and Plasma Substitutes of the National Research Council met and suggested that the results of the studies on Australia antigen should be reviewed in respect to eventually developing a test for donors. They did not, however, recommend specific action for testing. It was during 1969 that we began to hear of cases of pending or actual litigation in several parts of the country which had very clear effects on the introduction of testing in hospitals and blood banks. In several cases patients who had developed post-transfusion hepatitis, or their families, had sued hospitals and blood banks claiming that they had failed to test for Australia antigen. It was our impression that these legal actions effectively focused the interest of hospital authorities on the use of the Australia antigen test.

On July 18, 1970 Paul Schmidt and his colleague from the blood bank of the National Institutes of Health published an article in the "Point of View" section of Lancet entitled "Hepatitis associated antigen: To test or not to test?" (15) Using essentially the same data available to the National Research Council panel, they recommended that all blood bank laboratories equipped to do the test should do so; that is, a recommendation quite different from that of the NRC. The article contained a disclaimer from the National Institutes of Health to the effect that "This article was written by the authors in their private capacity. No official support or endorsement by the National Institutes of Health is intended or inferred." This was an unusual statement by the NIH which usually would stand by the publications of its scientific staff. Some ten days later, on July 29, 1970, a column appeared in the New York Times reporting on the Schmidt article and thus focused greater attention on the testing question.

The "To test or not to test" article, according to Dr. Schmidt, occasioned a reconvening of the NRC Committee on October 5, 1970. At this meeting they recommended that testing should be done, if possible. The official report was prepared rapidly and presented at the annual meeting of the American Association of Blood Banks in San Francisco on October 29, 1970, and published soon thereafter (16). A report was also made at that time by J. Stengle of the National Heart and Lung Institute of the NIH, that his Institute would make the test materials available to laboratories and blood banks. Hence, October 1970 may be taken as the date when many blood banks began to use the blood donor test for hepatitis B carriers. It required only 16 months between the time we became convinced of the validity of the data and many others did.

A variety of statutory events occurred after that. In 1971, legislation was passed in New York State requiring testing. (Legislation had also been enacted in other jurisdictions.) In the same year two companies, Spectra and Abbott, were licensed by the federal government to sell diagnostic kits. The American Association of Blood Banks, in their standards book of June 1972, ordained the use for testing. In July 1972 the Federal Register required testing for all blood banks involved in interstate practice, and the following year testing for all blood banks was mandated.

The identification of a hepatitis B virus was made in 1966 and by 1969 an immunological diagnostic test became routine; a rapid application of a basic research finding which started out as an esoteric investigation of inherited human variation. It could provide a good example of the vaslue of the support of basic research for the furtherance of clinical care.

USE IN RENAL DIALYSIS UNITS

The "Au test," as it was originally called, was also used to monitor renal dialysis units. These had begun to proliferate around the country following the passage of federal laws which provided financial support for those requiring dialysis. There had been major problems with hepatitis in the patients on these units; for the most part these infections had been chronic and often asymptomatic. In many cases, however, the staff of renal dialysis units acquired acute hepatitis which was often severe and in several cases fatal. We tested all patients and staff in a group of renal dialysis units in the Delaware Valley bimonthly for HBsAg and other markers. Similar procedures were followed in many other locations. By epidemiologic control alone and without (at that time) the use of gamma globulin or vaccine, it was possible to control hepatitis to such an extent that it was no longer a major hazard. Many of the dialysis units became and remained free of HBV infection.

The efficacy of the blood donor and institutional monitoring programs was significantly improved by the development of very sensitive methods for the detection of HBsAg. In 1970 we and other laboratories published methods for the application of the radioimmunoassay developed by Berson and Yallow, and these were quickly applied by commercial organizations who soon provided standardized test materials on a worldwide basis. These products had a significant impact on the acceptance of HBV testing. The test for HBsAg is now the most widely used radioimmunoassay in clinical chemistry.

GENETICS OF HBV

Our original entry into the hepatitis problem had been through population genetics. In the first extended paper on Australia antigen we had generated the hypothesis that the Australia antigen (or, as it was later understood, the carrier state for HBV) was under genetic control (4), and a variety of studies had been done after that to test and retest the hypothesis (17,18). We eventually concluded that there was genetic control of the mechanism by which the host responded to HBV infection, but that it could not be tested or even conceptualized by the classical means of family segregation analysis which we had utilized. We thought that the increasing knowledge of the molecular biology of HBV and its host could contribute to an understanding of this problem and deferred additional genetic studies until a later time. We had suggested that HBV DNA could enter the genome of a host. In some cases the egg and/or sperm cells might be infected in this manner and a gene related to host response to HBV could then segregate in families according to Mendalian predictions. It may be possible to directly test this model as the knowledge of the molecular biology of HBV increases.

SEX DIFFERENCES IN RESPONSE TO HBV INFECTION

Our studies of genetics have focused attention on families and sex difference, and this has led to some interesting observations and concepts. In our earliest studies we found that there appeared to be more male than female carriers in populations of apparently normal people, as well as in the disease groups in whom HBV was found in high frequency (17). This was confirmed in extensive population studies in Oceania, India, New Guinea, Surinam and in many other locations (19). London and Drew found that in a renal dialysis unit, males, when infected, were much more likely to develop the carrier state, and females, when infected, to develop antibody against the surface antigen (20); and similar studies from other places confirmed the striking sex differences.

We also looked at the effect of the virus on mothers and their offspring. We wanted to know how hepatitis B infection during pregnancy might affect the course of the pregnancy and/or the fetus. To that end, we studied over 800 pregnant women and 700 of their babies born at the Holy Family Hospital in New Delhi, India (21). By the use of the immunodiffusion technique, 7 HBsAg(+) pregnant women without clinical or laboratory signs of hepatitis were identified. The offspring of these women were all healthy and without congenital anomalies. Specifically, they did not differ from the children born to HBsAg(-) mothers with respect to length, weight, or head circumference. The HBsAg(+) mothers, however, did differ significantly from HBsAg(-) mothers in two respects: they were younger and had longer gestation times. In addition, there was a suggestion of reduced fertility. None of the HBsAg(+) women (by history) had been pregnant more than three times, whereas 20% of the HBsAg(-) controls had had three or more children. We suggested in the report of that study that the findings of younger age and fewer children among carrier women might represent either an effect of shorter time in the child-bearing age or relative infertility of older HBsAg(+) women. Although the differences we found were small, these findings encouraged us to conduct other investigations.

In earlier studies in collaboration with J. Friedlaender, we had investigated the familial segregation of the carrier state (HBsAg) on the island of Bougainville, Papua New Guinea. The data from this study were not available in a form which would allow formal analyses of fertility or sex ratio of the offspring, but they did generate the hypothesis that fertility and/or sex ratio would be different in families dependent on the HBV status of parents (i.e. if they were carriers or if they had developed anti-body to the surface antigen, anti-HBs). These hypotheses were tested in a field study carried out in 1973 by Hesser and her colleagues (22) in Plati, a village in Greek Macedonia. A significant increase in the secondary sex ratio (number of male/number of female live births X 100) was observed if either parent was HBsAg(+) (sex ratio = 185) compared with the offspring of couples in which both parents were HBsAg(-) (sex ratio = 113). In the initial analyses significant differences in family size were not noted.

Later, after we had observed that transplant patients who were anti-HBs(+) prior to kidney transplantation had survival experiences significantly different from chronic carriers (to be discussed later), we reanalyzed the data from Plati. In the second analysis carried out by Drew et al. (23), the families were divided into three groups:

1. Families in which either parent was HBsAg(+):anti-HBs(-)

2. Families in which both parents were HBsAg(-):anti-HBs(-)

3. Families in which both parents were HBsAg(-) and either parent was anti-HBs(+)

This categorization revealed a very high sex ratio (250) in the offspring of HBsAg(+) parents, an intermediate sex ratio (146) in the children of the HBsAg(-), anti-HBs(-) parents, and the lowest sex ratio in the offspring of the anti-HBs(+) parents (109). Further analysis showed that anti-HBs in mothers, but not in fathers, was associated with the lower sex ratio. Other variables thought to influence secondary sex ratio were examined by both Hesser and Drew. Parental age, total number of pregnancies, birth order, child-to-parent transmission of HBV, and socioeconomic factors were evaluated and could not account for the observed sex ratio differences.

Similar studies have now been completed on the island of Kar Kar off the coast of New Guinea (24) and in two communities in Greenland (unpublished). In these populations secondary sex ratios are also associated with the HBV responses of their parents, and the sex ratios are similar to those observed in Greece.

There may also be a fertility effect (25). In both Plati and Kar Kar, the number of sons born to each mating category (carrier parents or antibody positive mothers) was about the same (1.6-1.8 in Plati, 2.0-2.6 in Kar Kar), but the number of daughters per couple differed significantly (0.7 in HBsAg(+) couples, 1.4 in anti-HBs(+) couples in Plati; 2.3 in HBsAg(+) and 3.2 in anti-HBs(+) couples in Kar Kar). This could be explained by a combination of biological and behavioral factors affecting family size. We suggested that in Greece, where a preference for sons is known to exist, couples would continue to have children until they had the desired number of sons. Because of the effect on sex ratio of the parents' HBV responses, HBsAg(+) couples would achieve this end with fewer daughters, whereas anti-HBs(+) couples would have to have more daughters and more children altogether to produce the desired number of boys. Alternatively, HBV transmitted from a carrier parent might replicate more rapidly in females and therefore be more lethal to female than male embryos. This would result in the increased secondary sex ratios of HBV carrier couples which have been observed. Mothers with anti-HBs would protect embryos from HBV infection, thus eliminating selection pressure on female fetuses. Additional family studies have now been completed in the community of Matas Na Kahoy in Luzon, Philippines (unpublished) and these have confirmed the earlier studies. Hence, in all the populations carrier parents have fewer children, and this is due primarily to a decrease in the number of females.

Taken together, it appears from these studies that HBV has a profound effect on fertility and sex ratio. This, in addition to the direct effect of the virus on the mortality of patients and on the offspring of carriers, constitutes a major demographic factor which should be taken into account when worldwide prevention programs for HBV are undertaken.

RENAL TRANSPLANTATION

London and his colleagues in our laboratory have shown that the HBV response of renal patients has a bearing on the survival of transplanted kidneys (26). The transplanted kidneys of patients who have anti-HBs prior to transplantation have a shorter survival than those of patients who are carriers of HBsAg or who have no evidence of infection with HBV. This difference is particularly striking if the transplanted kidney is from a male; that is, anti-HBs appears to have an anti-male tissue effect.

They then developed a cellular model to explain the three kinds of sex differences discussed in this section; namely, 1) Males are more likely to become carriers and females are more likely to develop antibody when infected with HBV, 2) "Carrier" parents have a higher sex ratio of offspring than "anti-HBs" parents, 3) Renal patients with anti-HBs are more likely to reject transplanted male kidneys than patients without anti-HBs.

These three observations could all be explained by the hypothesis that there is an antigen on the cells of some males which cross-reacts with HBsAg. This cross-reactivity would make males more likely to recognize HBsAg as "self" and therefore tolerate the antigen (i.e., become chronic carriers). Conversely, females would be more likely to recognize HBsAg as "non-self" and produce anti-HBs.

In kidney graft recipients, tolerance to HBsAg would result in tolerance of male tissues and thus long-term survival of grafts from male donors in carrier recipients. Anti-HBs in a recipient, however, would react with male antigens on allografts resulting in early rejection of grafts from male donors. We can speculate that tolerance to HBsAg in potential mothers would result in lack of sensitization against male tissues and better survival of male fetuses. Anti-HBs would react with male tissues and thus hinder fertilization by Y-bearing sperm, interfere with implantation of male embryos or increase the probability of spontaneous abortion of male fetuses. Finally, HBsAg(+) carrier fathers would have HBsAg in their semen, which conceivably could protect Y-bearing sperm from anti-HBs or other cross-reacting ("anti-male") antibodies in their spouses' reproductive tracts.

Hence, the study of male/female differences has resulted in an interesting testable cellular model with a consequent greater insight into the interaction of HBV and human cells, as well as data which may be necessary for population planning programs if the use of the vaccine becomes widespread.

INSECT TRANSMISSION

The study of insect transmission of HBV can be of both practical and theoretical interest. If insect transmission is important, then insect control may be of great importance in public health programs. There are many interesting questions of a more hypothetical nature. Does replication occur in insects? If so, would there be differences between viruses which have replicated in insects and those which have replicated in humans in respect to surface and/or core antigens and in the DNA structure? Does the mode of transmission affect the host response? Are people differently affected if the virus infects them through an insect vector rather than through, say, blood transfusion?

By 1970 several authors had suggested that mosquito transmission might play a role in the dissemination of HBV, and Prince and his colleagues reported HBV in mosquitoes in 1972 (27). We studied mosquitoes collected in Uganda, Ethiopia and later Senegal and found field infection rates of 1 in 100 in several species collected in the wild, a very high rate (28,29). An even higher rate of infection, more than 50%, was found in bedbugs (*Cimex hemipterus*) collected from beds whose main occupants were carriers of HBV. Ogston and his colleagues in a series of studies showed that *Cimex* contains detectable HBsAg for up to six weeks (but in most bugs only four weeks) after a single feeding of blood containing infectious HBV (30). They later showed that HBV persisted in the feces up to eight weeks after feeding had commenced (31). Hence, the bedbugs may contain HBV material for a very long time. It has not been shown that the virus can actually be transmitted by the insects (the appropriate animal experiments have not been systematically done), but the possibility exists that they may be an important mechanism for mass transmission. Ogston further suggested that the dust of the feces inhaled or absorbed into open wounds might be a source of infection in addition to the more conventional method of biting and blood exchange. A scanning electron micrograph of a *Cimex* is shown in Figure 4.

HBV AND THE PREVENTION OF PRIMARY CANCER OF THE LIVER

It is likely that the most important consequence of the discovery of hepatitis B virus will be the understanding of its role in the prevention (and, possibly, the treatment) of one of the most common cancers in the world, primary cancer of the liver (also referred to as primary hepatocellular carcinoma, PHC). If this proves to be true, as appears likely, then it will be the first example of a common human cancer caused by a virus and for which a vaccine is available for its control. The HBV/PHC relation should also make a major contribution to the understanding of how viruses cause cancer in humans and help to formulate models

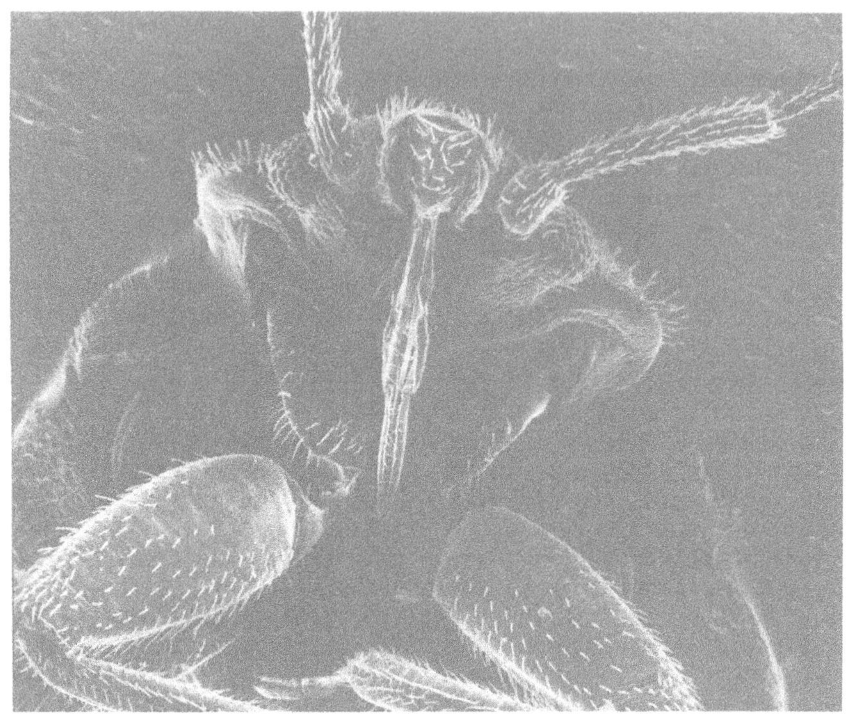

Figure 4. Scanning electron micrograph of head of bedbug Cimex hemipterus. Magnification approximately 100x. Electron micrograph courtesy of T. Anderson.

for prevention and control. (For a more comprehensive review, see references 32 and 33).

In sub-Saharan Africa, Taiwan, and the populous coastal and southern provinces of mainland China, the incidence of primary hepatocellular carcinoma is 25 to 150 cases per 100,000 population. An annual mortality of 100,000 for all of China (population base 850 million) has recently been reported. The incidence of PHC is 3 to 9 times higher in males than females. Hence, the estimated incidence of deaths from PHC in males would be about 17 to 20 per 100,000. Extrapolating these figures to other regions of the world where PHC is a common cancer, we can estimate a worldwide annual incidence of about 1/4 to 1 million cases in men and 50 to 200,000 cases in women. Since PHC is almost always lethal, the incidence and mortality rates per year are about the same (34).

In 1975 (35), we pointed out that advances in our knowledge of the pathology, epidemiology and clinical characteristics of PHC on the one hand, and of hepatitis B virus on the other, made it possible to test the hypothesis that persistent infection with the hepatitis B virus was necessary for the development of most cases of PHC. Since that time, a substantial body of evidence which strongly supports this hypothesis has been collected, and this will be briefly reviewed here.

1) PHC occurs commonly in regions where chronic carriers of HBV are common and much less frequently in areas where they are not.

2) Case-control studies have shown that as many as 90% or more of patients with PHC who live in areas where HBV is endemic have HBsAg or high titers of antibody against the core antigen in the blood (anti-HBc). These markers can be considered evidence of current or previous persistent HBV infection. In the same areas, controls have markedly lower frequencies of HBsAg and anti-HBc. Even in the United States, where PHC is uncommon, patients with the disease have higher prevalences of HBsAg and anti-HBc than do controls. In other words, in areas of both high and low PHC incidence, serologic evidence of persistent infection with HBV is more common in patients with PHC than in controls.

3) Most cases of PHC (approximately 80%) arise in a liver already affected with cirrhosis or chronic active hepatitis or both. If chronic hepatitis and cirrhosis are steps toward the development of liver cancer, then case-control studies of these two diseases should also show higher prevalence of chronic infection with HBV in the cases. Studies in Africa and Korea have confirmed this prediction.

4) HBV proteins can usually be demonstrated with histochemical stains or immunologic techniques in the hepatic tissues of patients with PHC. HBsAg and hepatitis B core antigen (HBcAg) are undetectable or present only in small quantities in the tumor cells themselves, but are found in the nonmalignant cells adjacent to the expanding tumor and elsewhere in the liver. These antigens are not found in the livers of uninfected persons nor in persons with antibody to HBsAg in their serum. A particularly pertinent study by Nayak et al. (36) in India, which is a relatively low incidence area for PHC, showed that if multiple sections of the liver are examined, HBsAg can be found in some hepatocytes in over 90% of the cases of PHC. Thus, HBV proteins are present in the livers of most patients with PHC from areas endemic for HBV and PHC, and they are also found in the livers of many patients from low incidence areas (36,37).

5) If persistent HBV infection causes PHC, such infection should precede the occurrence of PHC. To test this hypothesis, it is necessary to identify asymptomatic chronic carriers of HBV and

controls who are not carriers and to follow them for several years to see whether PHC develops. A major study of this type is being conducted by Beasley and his colleagues (38) in male civil servants between the ages of 40 and 60 years in Taiwan. Approximately 3500 carriers were identified. The controls are an equal number of HBsAg(-) men matched by age and place of origin. Approximately 18,500 additional non-carriers were also identified. The subjects have been followed for two to four years. Ninety-four cases of PHC have occurred during the follow-up period, and all but one have been in chronic carriers. The one exception arose in a man who had both anti-HBc and anti-HBs, indicating that he had probably been infected in the recent past. Thus far, the relative risk of PHC is more than 250 times greater in carriers than in non-carriers, and 98% (the attributable risk) of the cases have occurred in carriers. This is probably the highest risk known for any of the common cancers.

6) Because PHC usually develops in a liver that is affected by cirrhosis or chronic hepatitis or both, some investigators have argued that any hepatotoxic agent that causes cirrhosis is associated with an increased risk of PHC, and that hepatitis B virus is one such agent. A rigorous test of the hypothesis that chronic infection with hepatitis B virus increases the risk of PHC in addition to producing cirrhosis is to compare the incidence of PHC in patients with cirrhosis who are or are not chronic carriers of HBV. Obata et al. have performed such a study in Japan (39). Seven of 30 HBsAg(+) patients with cirrhosis (23%) but only five of 85 HBsAg(-) patients with cirrhosis (6%) had PHC after about four years. These results are highly consistent with the prediction from the hypothesis.

7) In populations where HBV is endemic (sub-Saharan Africa, Asia, Oceania, etc.), there is good evidence that many of the chronic carriers acquire HBV as a result of infection transmitted from their mothers early in life. (Although children of carrier mothers may be exposed _in utero_, at birth, or immediately afterwards, they do not become carriers until after about six weeks to three months.) That is, the mothers themselves are chronic carriers, and offspring born when the mothers are infectious are likely to become chronic carriers. Within a population, persons infected shortly after birth or during the first year of life will have been chronic carriers of HBV longer than persons of similar age who are infected later in life. Therefore, if the duration of being a chronic carrier is related to the likelihood of having PHC, one could predict that the mothers of patients with PHC would be more likely to be chronic carriers than the mothers of controls of similar age who do not have PHC. Studies in Senegal, West Africa (40) and in Korea (41) are consistent with this prediction.

8) A further test of the HBV-PHC hypothesis is whether HBV DNA is present in PHC tissue and whether such DNA is integrated

into the tumor cell genome. Summers et al. (42), using livers obtained at autopsy in Senegal, extracted HBV DNA base sequences from 9 of 11 primary liver cancers collected from patients with HBsAg in their serum and from one of four patients who were HBsAg(-) but anti-HBs(+) (42, 43). Several cell lines which produce HBsAg have been developed from human primary liver cancer. The first was produced by Alexander in South Africa (PLC/PRF/5) and has been studied in many laboratories (44). This cell line produces large quantities of HBsAg 22 nm particles (1.3 mg/ml) but no Dane particles (45). Marion and Robinson (46) analyzed these cells and demonstrated 4 to 5 ng of HBV DNA per mg of cellular DNA. Recently, Gray et al. (47) reported that at least six copies of HBV DNA are integrated into the cellular genomic DNA of the liver. Two of the six inserts are incomplete viral genomes, but all six contain the gene for HBsAg. They also isolated RNA transcripts for the HBsAg gene from the Alexander cell line. Brechot et al. (48) demonstrated integration of viral DNA in the cellular genomes of three primary liver cancers obtained at autopsy from HBsAg(+) men who lived in Ivory Coast, West Africa. Because only a few bands of integrated DNA were observed in each tumor extract, it is likely that the integration sites were the same within each cell of a given tumor. A third cell line that produces HBsAg has been derived from a human PHC by Aden and Knowles (49). Recently, Shafritz et al. (50) studied percutaneous liver biopsies and postmortem tissue specimens from patients with chronic liver disease associated with persistent HBV infection, and patients with PHC. In 12 patients with hepatocellular carcinoma who had persistent HBV in their serum, integrated HBV DNA was found in host liver cells. It was also found integrated in some patients who had PHC with anti-HBs. In addition, integration was found in the nontumorous tissue. In carriers of HBV without PHC, integration was seen in two patients who were carriers for more than eight years, but it was not integrated in individuals who were carriers for less than two years. From this it can be inferred that increasing time of infection increases the probability of integration.

9) In 1971 (51), based on the unusual clinical and epidemiologic characteristics of HBV, we had proposed that it was different from other viruses and that it represented the first of a series of viruses we termed Icrons (an acronym of the Institute for Cancer Research, ICR, with a neuter Greek ending). The unique characteristics of the molecular biology of HBV have supported the notion that HBV is an unusual virus. Recent discoveries of other viruses which conform to the expectations of the Icron model provide additional support for the hypothesis, and these will be briefly described here.

Persistent infection with a virus similar to HBV is associated with a naturally occurring primary carcinoma of the liver in *Marmota monax*, the woodchuck or groundhog (Figure 5). Robert

Figure 5. The eastern woodchuck or groundhog, *Marmota monax*. The photograph was taken at the woodchuck facility of the Division of Clinical Research at the New Bolton Center of the Veterinary School of the University of Pennsylvania. Photograph courtesy of L. Southam.

Snyder, Director of the Penrose Research Laboratory of the Philadelphia Zoological Garden, has trapped Pennsylvania woodchucks in the wild and maintained a colony at zoo for the past 20 years. Post-mortem examinations were performed in more than 100 woodchucks, and about 25% of the animals had primary liver cancers. The tumors in the animals were usually associated with chronic hepatitis (51). Summers, at the Institute for Cancer Research, examined serum samples from these animals for evidence of infection with a virus similar to HBV. He based his investigation on the hypothesis that viruses in the same class as HBV would have a similar nucleic acid structure and similar DNA polymerase. HBV was known to have unique characteristics: it contains a circular, double-stranded DNA genome with a single-stranded region and a DNA polymerase capable of filling in the single-stranded region to make a fully double-stranded, circular DNA. Summers found that about 15% of the woodchuck serum samples had particles containing a DNA polymerase and a DNA genome that were similar in size and structure to those of HBV (52). Examination of pellets from these serum specimens with an electron microscope showed the three types of particles associated with HBV. Later, Werner et al. (53) showed cross-reactivity of the core and surface antigens of the virus in woodchucks (WHV) with the comparable antigens of HBV. A close association between persistent WHV infection and PHC has also been found; DNA from WHV hybridized to the cellular DNA in five woodchuck livers containing PHC but did not hybridize to the DNA in nine livers without tumors. Finally, Summers and his colleagues have demonstrated integration of one or two WHV genomes into tumor-cell DNA in two woodchuck primary liver cancers. Integration appeared to occur at the same unique site in each cell of the tumor. Thus, each tumor was a clone with respect to the integrated viral DNA, a finding similar to that in the humans.

Liver cancer in the woodchuck is not what is generally regarded as a laboratory model of a human disease, that is, it was not designed or "created" by an investigator for research purposes; rather, it is a naturally occurring disease related to a naturally occurring virus, both of which have remarkable features in common with their human counterparts. It provides impressive support for the hypothesis and also an opportunity to perform observations and studies with an other-than-human species.

An "Icron" has been found in some breeds of Chinese domestic ducks (in which there is a high frequency of liver cancer) as well as domestic Pekin ducks in the United States. One has also been found in California ground squirrels (*Spermophilus beecheyi*) (54).

Our interpretation of these nine lines of evidence is that, taken together, they strongly support the hypothesis that persistent infection with HBV is required for the development of most cases of PHC, and therefore that the next step is warranted: testing of the hypothesis that decreasing the frequency of HBV infection will in due course decrease the frequency of PHC. The availability of the hepatitis B vaccine produced from HBsAg in human blood (10) and increasing knowledge of the mechanisms of transmission of HBV will make such a study feasible. Since the incidence of cancer is high in HBV carriers, it may be possible to measure the effect of the program within a reasonable time. In addition, the control of HBV infection is clearly justified as a public health measure for the prevention of acute and chronic hepatitis and postnecrotic cirrhosis, diseases of major importance in the same regions where PHC is common.

A MODEL FOR HEPATITIS B VIRUS AND PRIMARY CANCER OF THE LIVER

Beginning in 1980, London and Blumberg developed a model to explain how HBV "causes" PHC incorporating the clinical, epidemiologic and molecular observations pertaining to the relation, and some of the concepts which derive from our original interest in human variation. This model has been described in detail elsewhere (32) but will be reviewed briefly here.

The model posits the existence of two types of cells in respect to infection with HBV. One, the S cell, is a fully differentiated cell. When it is infected with HBV, it is susceptible to productive infection and many copies of the virus are replicated. The R cell is a less differentiated cell. When it is infected, integration of HBV DNA may occur, but the infection is nonproductive. The death of the productively infected S cells, which can be due to lysis of the cell, immune response by the host or other causes, stimulates the division of the R cell. These may divide to form two S cells to replace those that were killed, one S cell and one R cell, to allow the continuation of the replacement process, or two R cells. If uncontrolled production of R cells occurs, which may be due to some other "event," then a tumor of a size perceptible to the host may develop; that is, clinical cancer has occurred.

This model leads to some interesting preventative and therapeutic possibilities. Most forms of cancer therapy are directed towards the killing of cancer cells, often at the expense of normal cells. This model recommends other measures; for example, to preserve the health of the infected S cells, or to prevent the transfer of virus from one cell to another, or to inhibit the message which causes the division of R cells. The HBV infected people can be identified at a very young age, and we are seeking measures that would slow down the process whereby R cells accumulate so that the host can live out his or her life span before perceptible clinical cancer develops.

CONCLUSIONS

The original studies on Australia antigen have led to a wide variety of research findings with both practical and theoretical interest. It is likely that the next 20 years will produce even more interesting and, it is hoped, valuable results.

ACKNOWLEDGMENTS

This work was supported by USPHS grants CA-06551, RR-05539, CA-06927 and CA-22780 from the National Institutes of Health and by an appropriation from the Commonwealth of Pennsylvania.

REFERENCES

1. Allison, A. C., Blumberg, B. S., and Rees, A. P. (1958). Haptoglobin types in British, Spanish Basque and Nigerian African populations. Nature 181:824-825.
2. Blumberg, B. S., Allison, A. C., and Garry, B. (1959). The haptoglobins and hemoglobins of Alaskan Eskimos and Indians. Ann. Hum. Genet. 23:340-356.
3. Blumberg, B. S., Dray, S., and Robinson, J. C. (1962). Antigen polymorphism of a low-density beta-lipoprotein. Allotype in human serum. Nature 194:656-658.
4. Blumberg, B. S., Alter, H. J., and Visnich, S. (1965). A "new" antigen in leukemia sera. J. Am. Med. Assoc. 191:541-546.
5. Blumberg, B. S., Gerstley, B. J. S., Hungerford, D. A., London, W. T., and Sutnick, A. I. (1967). A serum antigen (Australia antigen) in Down's syndrome leukemia and hepatitis. Ann. Intern. Med. 66:924-931.
6. Sutnick, A. I., London, W. T., Gerstley, B. J. S., Cronlund, M. M., and Blumberg, B. S. (1968). Anicteric hepatitis associated with Australia antigen. Occurrence in patients with Down's syndrome. J. Amer. Med. Assoc. 205:670-674.
7. Bayer, M. E., Blumberg, B. S., and Werner, B. (1968). Particles associated with Australia antigen in the sera of patients with leukemia, Down's syndrome and hepatitis. Nature 218:1057-1059.
8. Dane, D. S., Cameron, C. H., and Briggs, M. (1970). Virus-like particles in serum of patients with Australia antigen-associated hepatitis. Lancet 1:695-698.
9. Okochi, K. and Murakami, S. (1968). Observations on Australia antigen in Japanese. Vox. Sang. 15:374-385.
10. Blumberg, B. S. and Millman, I. Vaccine against viral hepatitis and process. Serial No. 864,788 filed 10/8/69, Patent 36 36 191 issued 1/18/72, U.S. Patent Office.
11. Okochi, K., Murakami, S., Ninomiya, K., and Kaneko, M. (1970). Australia antigen, transfusion and hepatitis. Vox. Sang. 18:289-300.
12. Senior, J. R., Sutnick, A. I., Goeser, E., London, W. T., Dahlke, M. D., and Blumberg, B. S. (1974). Reduction of post-transfusion hepatitis by exclusion of Australia antigen from donor blood in an urban public hospital. Am. J. Med. Sci. 267:171-177.
13. Blumberg, B. S., London, W. T., and Sutnick, A. I. (1969). Relation of Australia antigen to virus of hepatitis. Bull. of Path. 10:164.
14. Goche, D. J. and Kavey, N. B. (1969). Hepatitis antigen: correlation of disease with infectivity of blood donors. Lancet 1:1055.

15. Alter, H. J., Holland, P. V., and Schmidt, P. J. (1970). Hepatitis-associated antigen: to test or not to test? Lancet 2:142-143.
16. Ad hoc committee on hepatitis-associated antigen. (1971). Statement concerning the use of hepatitis-associated antigen tests for donor screening in blood banks. Transfusion 11:1-3.
17. Blumberg, B. S., Melartin, L., Guinto, R. A., and Werner, B. (1966). Family studies of a human serum isoantigen system (Australia antigen). Am. J. Hum. Genet. 18: 594-608.
18. Blumberg, B. S., Friedlaender, J. S., Woodside, A., Sutnick, A. I., and London, W. T. (1969). Hepatitis and Australia antigen. Autosomal recessive inheritance of susceptibility to infection in humans. Proc. Natl. Acad. Sci. (USA) 62:1108-1115.
19. Blumberg, B. S., Sutnick, A. I., London, W. T., and Melartin, L. (1972). Sex distribution of Australia antigen. Arch. Intern. Med. 130:227-231.
20. London, W. T. and Drew, J. S. (1977). Sex differences in response to hepatitis B infection among patients receiving chronic dialysis treatment. Proc. Natl. Acad. Sci. (USA) 74:2561-2563.
21. Kukowski, K., London, W. T., Sutnick, A. I., Kahn, M., and Blumberg, B. S. (1972). Comparison of progeny of mothers with and without Australia antigen. Human Biol. 44:489-500.
22. Hesser, J. E., Economidou, J., and Blumberg, B. S. (1975). Hepatitis B surface antigen (Australia antigen) in parents and sex ratio of offspring in a Greek population. Human Biol. 47:415-425.
23. Drew, J. S., London, W. T., Lustbader, E. D., Hesser, J. E., and Blumberg, B. S. (1978). Hepatitis B virus and sex ratio of offspring. Science 201:687-692.
24. Drew, J., London, W. T., Blumberg, B. S., and Serjeantson, S. (1982). Hepatitis B virus and sex ratio on Kar Kar Island. Human Biol. 54:123-135.
25. London, W. T., Stevens, R. G., Shofer, F. S., Drew, J. S., Brunhofer, J. E., and Blumberg, B. S. (1982). Effects of hepatitis B virus on the mortality, fertility, and sex ratio of human populations, in Viral Hepatitis, 1981 International Symposium, W. Szmuness, H. J. Alter and J. E. Maynard, eds., Franklin Institute Press, Philadelphia, pp. 195-202.
26. London, W. T., Drew, J. S., Blumberg, B. S., Grossman, R. A., and Lyons, P. J. (1977). Association of graft survival with host response to hepatitis B infection in patients with kidney transplants. N. Engl. J. Med. 296: 241-244.

27. Prince, A. M., Metselaar, D., Kafuko, G. W., Mukwaya, L. G., Ling, C. M., and Overby, L. R. (1972). Hepatitis B antigen in wild-caught mosquitoes in Africa. Lancet 2: 247-250.
28. Blumberg, B. S., Wills, W., Millman, I., and London, W. T. (1973). Australia antigen in mosquitoes. Feeding experiments and field studies. Res. Commun. Chem. Path. Pharm. 6:719-732.
29. Wills, W., Saimot, G., Brochard, C., Blumberg, B. S., London, W. T., Dechene, R., and Millman, I. (1976). Hepatitis B surface antigen (Australia antigen) in mosquitoes collected in Senegal, West Africa. Am. J. Trop. Med. Hyg. 25:186-190.
30. Ogston, C. W., Wittenstein, F. S., London, W. T., and Millman, I. (1979). Persistence of hepatitis B surface antigen in the bedbug _Cimex hemipterus_ (Fabr.). J. Infect. Dis. 140:411-414.
31. Ogston, C. W., and London, W. T. (1980). Excretion of hepatitis B surface antigen by the bedbug _Cimex hemipterus_ Fabr. Trans. Roy. Soc. Trop. Med. Hyg. 74:823-825.
32. London, W. T. and Blumberg, B. S. (1982). Hepatitis B and related viruses in chronic hepatitis, cirrhosis and hepatocellular carcinoma in man and animals, _in_ Chronic Active Liver Disease, S. Cohen and R. D. Soloway, eds., Churchill Livingstone, New York, pp. 147-170.
33. Blumberg, B. S., and London, W. T. (1983). Hepatitis B virus pathogenesis and prevention of primary cancer of the liver, _in_ Accomplishments in Cancer Research 1981, General Motors Cancer Research Foundation, J. G. Fortner and J. E. Rhoads, eds., J. B. Lippincott Company, Philadelphia, pp. 184-201.
34. London, W. T. (1981). Primary hepatocellular carcinoma--etiology, pathogenesis, and prevention. Human Pathology 12:1085-1097.
35. Blumberg, B. S., Larouze, B., London, W. T., Werner, B., Hesser, J. E., Millman, I., Saimot, G., and Payet, M. (1975). The relation of infection with the hepatitis B agent to primary hepatic carcinoma. Am. J. Pathol. 81:669-682.
36. Nayak, N. C., Dhark, A., Sachdeva, R., Mittal, A., Seth, H. N., Sudarsanam, D., Reddy, B., Wagholikar, U. L., and Reddy, C. R. R. M. (1977). Association of human hepatocellular carcinoma and cirrhosis with hepatitis B virus surface and core antigens in the liver. Int. J. Cancer 20:643.
37. Omata, M., Ashcavi, M., Liew, C-T., and Peters, R. L. (1979). Hepatocellular carcinoma in the U.S.A., etiologic considerations: localization of hepatitis B antigens. Gastroenterology 76:279.

38. Beasley, R. P., Lin, C-C., Hwang, L-Y., and Chien, C-S. (1981). Hepatocellular carcinoma and hepatitis B virus: A prospective study of 22,707 men in Taiwan. Lancet 2:1129.
39. Obata, H., Hayashi, N., Motoike, Y., et al. (1980). A prospective study on the development of hepatocellular carcinoma from liver cirrhosis with persistent hepatitis B virus infection. Int. J. Cancer 25:741.
40. Larouze, B., London, W. T., Saimot, G., Werner, B. G., Lustbader, E. D., Payet, M., and Blumberg, B. S. (1976). Host responses to hepatitis B infection in patients with primary hepatic carcinoma and their families. A case/control study in Senegal, West Africa. Lancet 2:534-538.
41. Hann, H. L., Kim, C. Y., London, W. T., Whitford, P., and Blumberg, B. S. (1982). Hepatitis B virus and primary hepatocellular carcinoma: Family studies in Korea. Int. J. Cancer 30:47-51.
42. Summers, J., O'Connell, A., Maupas, P., Goudeau, A., Coursaget, P., and Drucker, J. (1978). Hepatitis B virus DNA in primary hepatocellular carcinoma tissue. J. Med. Virol. 2:207-214.
43. London, W. T. Observations on hepatoma, in Viral Hepatitis, G. N. Vyas, S. N. Cohen and R. Schmid, eds., Franklin Institute Press, Philadelphia, pp.455-458.
44. Macnab, G. M., Alexander, J. J., Lecatsas, G., Bey, E. M., and Urbanowicz, J. M. Hepatitis B surface antigen produced by a human hepatoma cell line. Br. J. Cancer 34:509.
45. Skelly, J., Copeland, J. A., Howard, C. R., and Zuckerman, A. J. (1979). Hepatitis B surface antigen produced by a human hepatoma cell line. Nature 282:617.
46. Marion, P. L. and Robinson, W. S. (1980). Hepatitis B virus and hepatocellular carcinoma, in Viruses in Naturally Occurring Cancers, M. Essex, G. Todaro and H. zur Hausen, eds., Cold Spring Harbor Conferences on Cell Proliferation, Vol. 7, Cold Spring Harbor: Cold Spring Harbor Laboratory, pp. 423.
47. Gray, P., Edman, J. C., Valenzuela, P., Goodman, H. M., and Rutter, W. J. (1980). A human hepatoma cell line contains hepatitis B DNA and RNA sequences. J. Supramolec. Struc. (Suppl). 4:245.
48. Brechot, C., Pourcel, C., Louise, A., Rain, B., and Tiollais, P. (1980). Presence of integrated hepatitis B virus DNA sequences in cellular DNA of human hepatocellular carcinoma. Nature 286:533.
49. Knowles, B. B., Howe, C. C., and Aden, D. P. (1980). Human hepatocellular carcinoma cell lines secrete the major plasma proteins and hepatitis B surface antigen. Science 209:497.

50. Shafritz, D. A., Shouval, D., Sherman, H. I., Hadziyannis, S. J., and Kew, M. C. (1981). Integration of hepatitis B virus DNA into the genome of liver cells in chronic liver disease and hepatocellular carcinoma. N. Engl. J. Med. 305:1067.
51. Blumberg, B. S., Millman, I., Sutnick, A. I., and London, W. T. (1971). The nature of Australia antigen and its relation to antigen-antibody complex formation. J. Exp. Med. 134:320-329.
52. Summers, J., Smolec, J. M., and Snyder, R. (1978). A virus similar to human hepatitis B virus associated with hepatitis and hepatoma in woodchucks. Proc. Natl. Acad. Sci. USA 75:4533.
53. Werner, B. G., Smolec, J. M., Snyder, R., and Summers, J. (1979). Serological relationship of woodchuck hepatitis virus to human hepatitis B virus. J. Virol. 32:314-322.
54. Blumberg, B. S. (1981). Viruses similar to hepatitis B virus (Icrons). Human Pathology 12:1107-1113.

HEPATITIS B VIRUSES

Jesse Summers

Institute for Cancer Research
Fox Chase Cancer Center
Philadelphia, Pennsylvania 19111

VIRION AND GENOMIC STRUCTURE

Hepatitis B virus is a small DNA-containing virus that causes persistent noncytopathic infections of the liver. Infected hepatocytes continually secrete viral specific particles that accumulate to high levels (10^{13}/ml) in the blood. These particles (Figure 1) are of two types: i) noninfectious particles consisting of excess viral coat protein (HBsAg) and containing no nucleic acid, and ii) lower amounts (10^{10}/ml) of infectious, DNA-containing particles (Dane particles) consisting of a 27 nm nucleocapsid core (HBcAg) around which is assembled an envelope containing the major viral coat protein, carbohydrate, and lipid. The DNA genome is about 3000 nucleotides in length, is circular (1) and partly single stranded, containing an incomplete plus strand (2) (Figure 1). The incomplete plus strand is complexed with a DNA polymerase in the virion which, under appropriate *in vitro* conditions, can elongate it using the complete minus strand as the template. These morphological and structural features distinguish hepatitis B viruses from all known classes of DNA-containing viruses.

THE REPLICATION CYCLE

The replication cycle of hepatitis B viruses is strikingly different from other DNA-containing viruses and suggests a close relationship with the RNA-containing retroviruses. The main unusual feature is the use of an RNA copy of the genome as an intermediate in the replication of the DNA genome (Figure 2) (3). Briefly, infecting DNA genomes are converted to a double-stranded form(s) which serves as a template for transcription of RNA.

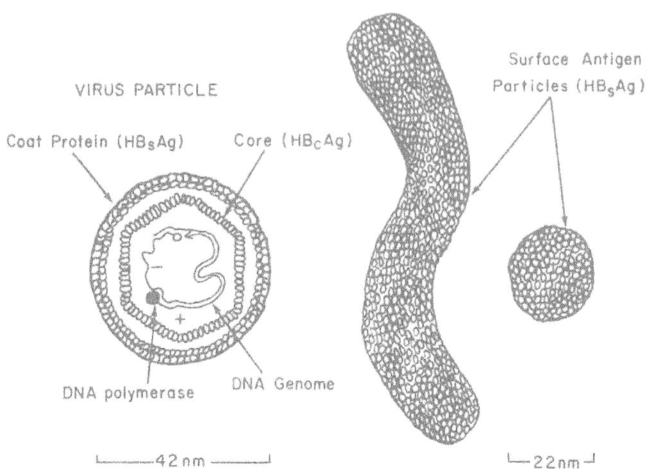

Figure 1. Virus specific particles and the genome structure of hepatitis B virus. Both infectious and noninfectious particles are produced by infected hepatocytes. Infectious particles contain a small circular DNA genome and a DNA polymerase.

Multiple RNA transcripts are synthesized from each infecting genome, and these transcripts either have messenger function or DNA replicative function. The latter, termed "pre-genomes," are precursors of the progeny DNA genomes because they are assembled into nucleocapsid cores and reverse-transcribed into DNA before coating and export from the cell. Thus each mature virion contains a DNA copy of the RNA pre-genome and a DNA polymerase.

The specific mechanism for reverse-transcription of the pre-genome, as well as the structure of the pre-genome, has not been entirely worked out. The pre-genome appears to be a plus strand RNA of approximately one genome length, as determined by sucrose gradient sedimentation and gel electrophoresis. Some aspects of the mechanism of reverse-transcription have been inferred through the structures of intermediates in this process. The first DNA to be synthesized is of minus strand polarity and

⟶

Figure 2. Proposed replication cycle of the hepatitis B viruses. Infecting DNA genomes are converted to double-stranded templates for transcription by a cellular RNA polymerase. This "proviral" DNA form(s) is responsible for the production of mRNA and pre-genomic RNA's. Pre-genomic RNA's are packaged into nucleocapsid cores, reverse-transcribed into DNA and matured into enveloped virions.

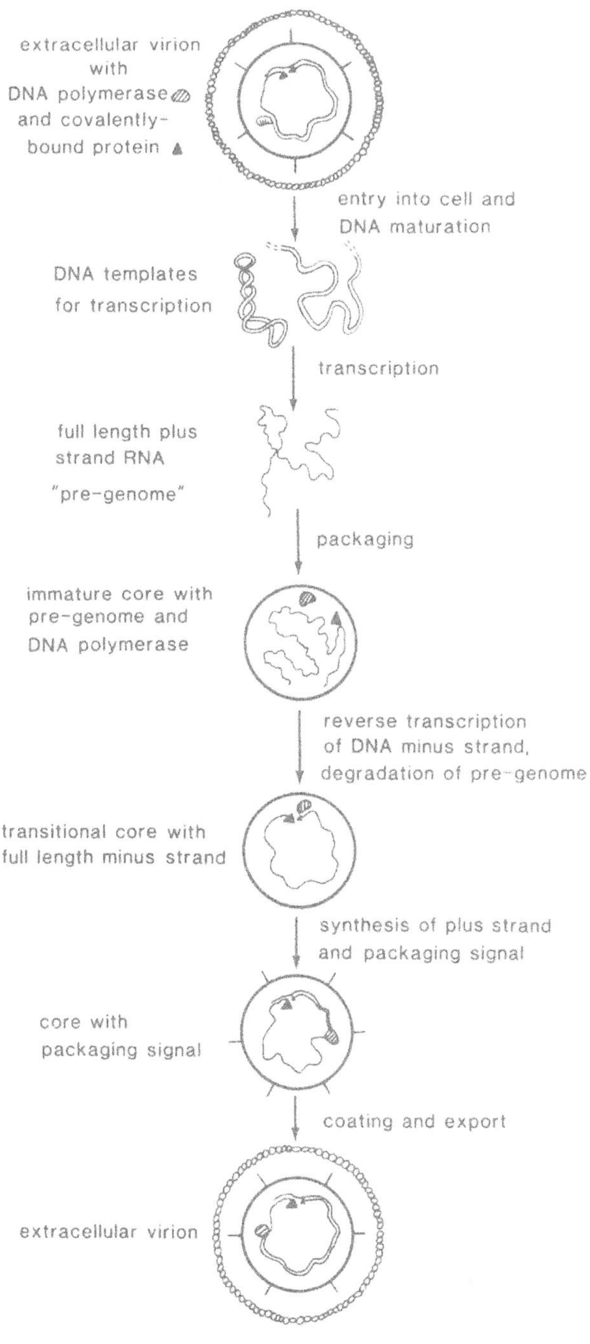
extracellular virion
with
DNA polymerase
and covalently-
bound protein
entry into cell and
DNA maturation
DNA templates
for transcription
transcription
full length plus
strand RNA
"pre-genome"
packaging
immature core with
pre-genome and
DNA polymerase
reverse transcription
of DNA minus strand,
degradation of pre-genome
transitional core with
full length minus strand
synthesis of plus strand
and packaging signal
core with
packaging signal
coating and export
extracellular virion

is initiated at a unique site on the viral genetic map. (See Figure 4). Very small (< 30 nucleotides) nascent DNA minus strands are covalently linked to a protein, and it is likely that this protein acts as the primer for minus strand DNA synthesis (4). Growth of the minus strand DNA is accompanied by a coordinate degradation of the pre-genome so that the product is a full-length single-stranded DNA, rather than an RNA:DNA hybrid (Figure 3). Plus strand DNA synthesis has been observed only after completion of the minus strand, and initiates at a unique site close to the 5' end of the minus strand (Figures 3 & 4). Complete elongation of the plus strand is not a requirement for coating and export of the nucleocapsid cores, thus most extracellular virions contain incomplete plus strands and a large single-stranded gap in the genome.

GENETIC ORGANIZATION OF HBV'S

The open reading frames present in the nucleotide sequences of cloned DNA's from HBV's have revealed an interesting and complex genetic organization. (5). Four genes are readily identifiable as long sequences beginning at an ATG and uninterrupted by termination codons. (Figure 4). These genes are encoded in the same strand of DNA, the minus strand, and their messages are

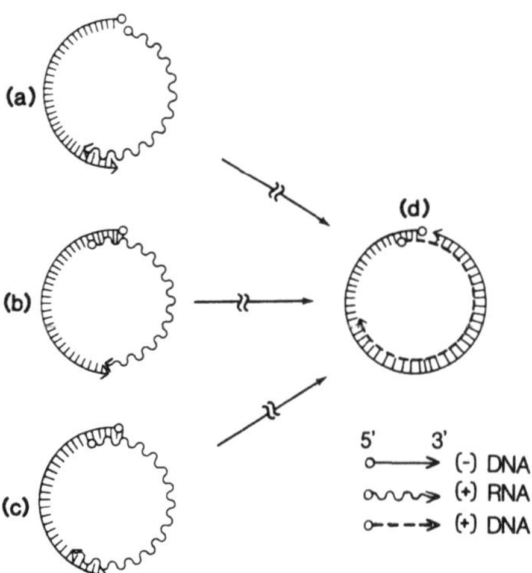

Figure 3. Possible structure of intermediates in the synthesis of HBV DNA. Nascent DNA minus strands are base paired with a plus strand RNA template at the growing 3' end of the minus strand (a), the site of initiation of DNA synthesis (b), or at both sites (c). The RNA template is degraded as it is transcribed into DNA. Plus strand DNA synthesis takes place on full-length minus strands, resulting in circularization of the genome.

therefore of the same polarity as the pre-genome, i.e. of plus strand polarity. This observation may suggest a single transcriptional unit for all RNA transcripts, i.e. a single promoter. No experimental evidence exists at the present time to confirm or exclude this hypothesis.

Each open reading frame overlaps at least one other open reading frame by a substantial amount. Thus all three registers are utilized in an amino acid coding capacity. The consequences of this organization to gene expression and genetic variation are not completely understood.

THE HBV GROUP, COMPARATIVE PATHOGENESIS

Four examples of hepatitis B viruses infecting man and other animals are known (reviewed in 6) (Table 1). These viruses, originally detected by serum screening, have been found to be like the human HBV, hepatotropic, and capable of long, persistent, productive infections. Two members of the group, HBV and WHV, cause serious chronic liver disease in the persistently infected hosts, while infections with GSHV and DHBV have very little pathological consequence. This finding is consistent with the view that viral-associated liver disease has an immunopathological basis. Thus, a

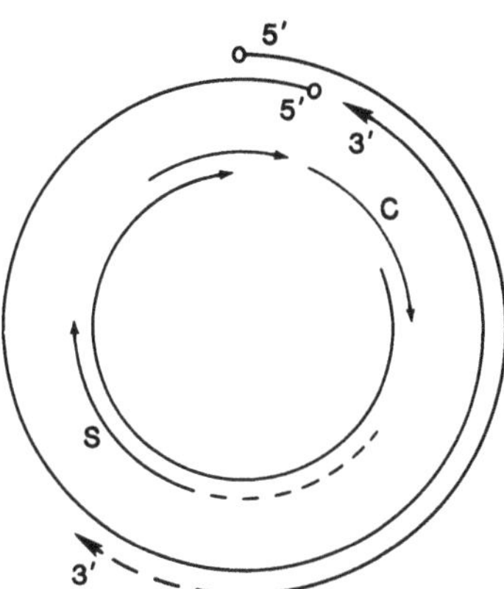

Figure 4. Genetic organization of HBV's. Derived from an analysis of the nucleotide sequence of HBV and WHV (4), the diagram shows that regions coding for polypeptides of the surface antigen (s), the core antigen (c) and two other polypeptides of unknown function. The complete minus strand and nascent plus strand of the DNA genome are shown along with the approximate sites of initiation of DNA synthesis (the 5' ends of the two strands).

Table 1. A Comparison of the Properties of Four Hepatitis B Viruses

	HBV	WHV	GSHV	DHBV
virions	42 nm spherical 27 nm core p = 1.24 in CsCl DNA polymerase activity	45 nm spherical 27 nm core cross-reactive with HBcAg (10%) p = 1.225 in CsCl DNA polymerase activity	47 nm spherical approx. 30 nm core cross reactive with HBcAg DNA polymerase activity	40-45 nm spherical 27 nm core (spikes) p = 1.16 in CsCl DNA polymerase activity
genome	DNA circular large single-stranded gap cohesive ends 3,182 base pairs	DNA circular large single-stranded gap cohesive ends 3,308 base pairs some homology with HBV	DNA circular large single-stranded gap cohesive ends 3,250-3,300 base pairs	DNA circular large single-stranded gap cohesive ends approx. 3,000 base pairs
"surface antigen" particles	"HBsAg" numerous in the blood 22 nm spherical and filamentous forms p = 1.19-1.20 in CsCl	"WHsAg" numerous in the blood 20-25 nm spherical and filamentous forms p = 1.18 in CsCl weak cross-reaction with HBsAg (0.1-1%)	"GSHsAg" numerous in the blood 15-25 nm spherical and long filamentous forms p = 1.18 in CsCl weak cross-reaction with HBsAg	"DHBsAg" numerous in the blood 40-60 nm spherical convoluted forms p = 1.14 in CsCl
"natural" host	human	eastern woodchuck (*Marmota monax monax*)	Beechey ground squirrel (*Spermophilus beecheyi*)	Pekin duck and occasionally other breeds (*Anas domesticus*)
distribution in selected populations	0.1-20% persistent infections	16-30% persistent infections	0-50% persistent infections	5-10% persistent infections
transmission	vertical horizontal	?	?	egg transmitted
tissue tropism	liver	liver	liver	liver, pancreas
associated disease	healthy carriers acute, chronic forms of hepatitis hepatocellular carcinoma	healthy carriers chronic forms of hepatitis hepatocellular carcinoma	healthy carriers ? ?	healthy carriers ? ?

genetically determined immunological capability of the host probably influences i) the course of an infection; i.e. acute or persistent, and ii) if persistent, the occurrence of chronic liver disease associated with that infection. Beechey ground squirrels, for example, in some study areas, are persistently infected with GSHV at frequencies approaching 50%. Their apparent immunological inability to resist persistent GSHV infections may be related to the lack of immunopathogenesis of the infection.

An additional consequence of infection with hepatitis B viruses is a high risk of hepatocellular carcinoma. Hepatocellular carcinoma is a major fatal disease in many areas of the world, and the vast majority of these cases can be attributed to chronic hepatitis B infections. In one study of hepatocellular carcinoma associated with human HBV infections, the cumulative risk for development of liver cancer was about 2% after 4.5 years of persistent infection (7). In WHV-infected woodchucks, the cumulative risk is around 80% after about four years of infection. Hepatocellular carcinoma associated with GSHV or DHBV infection has not been demonstrated. Considering the animal systems so far known, the risk of liver cancer correlates with the occurrence of other chronic liver disease, and thus hepatocellular carcinoma may also have an immunopathologic component.

Both human and woodchuck hepatocellular carcinomas commonly contain integrated viral DNA (8) (9) (10), indicating that the tumors were clonally derived from a virally infected cell. Such integrated viral DNA might function in tumorigenesis by activating a cellular oncogene, or by expressing a transforming viral protein. The available information does not confirm or exclude either of these mechanisms. Two recombinant DNA clones of integrated WHV DNA's from tumors containing a single site in each cell at which viral DNA was integrated have been analyzed (10). The viral sequences in these tumors were highly rearranged relative to virion DNA and contained only one viral gene that was present in its entirety (Figure 5). Further studies are required to determine whether integrated viral DNA functions in tumors, or whether viral integrations occur in common regions of the tumor cell genome.

VIRAL DNA'S IN PERSISTENT INFECTIONS

Integrated hepatitis viral DNA is not unique to hepatocellular carcinomas. HBV and WHV DNA's have been shown to be integrated in small amounts in infected liver tissue (11,12). Such integrated DNA's almost certainly function to produce viral specific proteins. In long-term HBV infections viral replication gradually subsides in the liver until neither virus production nor

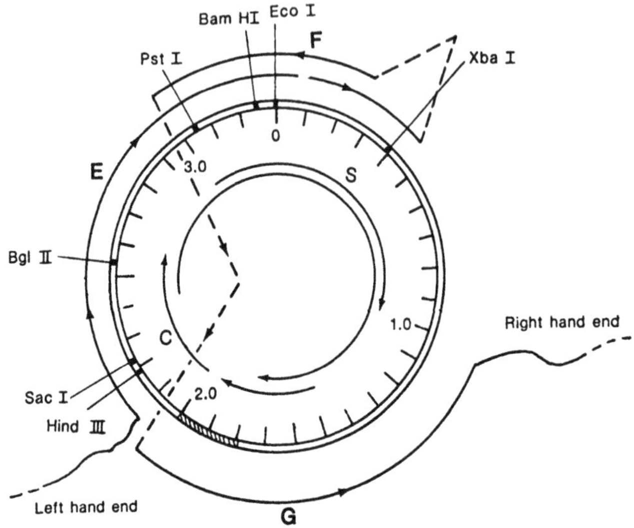

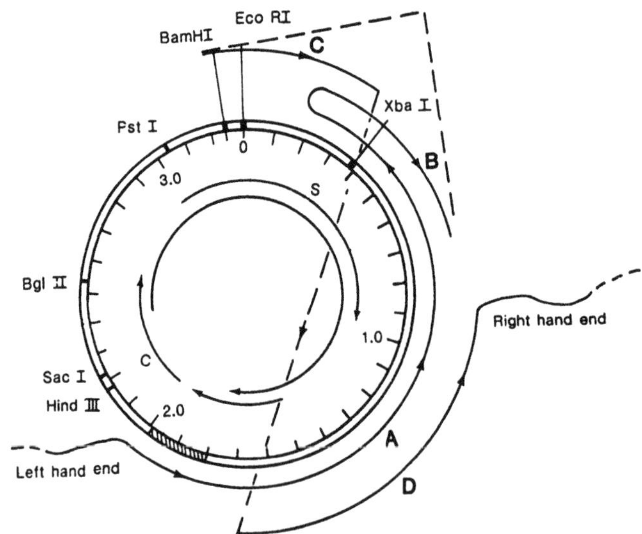

Figure 5. Genetic structure of two integrated woodchuck hepatitis virus genomes cloned from WHV-induced hepatocellular carcinomas. These cloned DNA's represent the only site at which viral DNA was integrated in the cells of the two respective tumors. The viral genetic map is shown on the inner circle, along with the positions of the open reading frames. The arrangement of viral sequences in the integrated DNA is indicated by the solid lines plotted along the outside of the genetic map. The viral sequences are highly rearranged.

free viral DNA's can be demonstrated in the liver. Hepatocytes nevertheless continue to synthesize and secrete viral coat protein in the form of surface antigen particles. In such livers, only integrated forms of viral DNA are detectable.

The gradual decline in virus production during chronic infections is accompanied by the appearance in the liver of increasing numbers of uninfected hepatocytes. Many of these cells had previously been infected, since high levels of integrated viral DNA are maintained in the liver in spite of the accumulation of these new cells. The data suggest a process by which infected cells are "cured," but nothing is known concerning such a process.

From the replication cycle (Figure 2), it can be seen that the ability of an infected cell to produce virus will depend on the continuous presence of a transcriptionally active form viral DNA. Such "proviral" DNA must be continuously present since each pregenomic RNA molecule can give rise to only one virus particle. The structure of proviral DNA is not known, nor is the mechanism by which it is maintained in infected cells. Integration of proviral DNA molecules into cellular DNA might destroy their ability to support virus production, i.e. to serve as templates for pregenomic RNA. Such a process might explain both "curing" of infected cells, and the presence of integrated viral DNA in "cured" cells. Other possible mechanisms that have been proposed to explain the gradual turnoff of viral synthesis involve massive cell turnover and selection for resistant cells (13).

SUMMARY

Hepatitis B viruses are noncytopathic in man and animals and replicate via a pathway similar to that of retroviruses. Persistent virus production at the cellular level depends on transcription of a stable form of double-stranded viral DNA (the provirus). Persistent infection results in the gradual accumulation of integrated viral DNA and "cured" cells in the infected liver. Tumors arising in persistently infected liver contain integrated viral genomes, but it is not known if these viral DNA's function directly in the oncogenic process.

ACKNOWLEDGMENTS

This work was supported by NIH grants AI-15166, CA-06927 and by an appropriation from the Commonwealth of Pennsylvania.

REFERENCES

1. Robinson, W. S., Clayton, D. A., and Greenman, R. L. (1974). DNA of a human hepatitis B virus candidate. J. Virol. 14:384-391.
2. Summers, J., O'Connell, A., and Millman, I. (1975). Genome of hepatitis B virus: restriction enzyme cleavage and structure of the DNA isolated from Dane particles. Proc. Natl. Acad. Sci. (USA) 72:4597-4601.
3. Summers, J. and Mason, W. S. (1982). Replication of the genome of a hepatitis B-like virus by reverse transcription of an RNA intermediate. Cell 29:403-415.
4. Molnar-Kimber, K., Summers, J., Taylor, J. M., and Mason, W. S. (1983). Protein covalently bound to Minus-strand DNA intermediates of duck hepatitis B virus. J. Virol. 45:165-172.
5. Galibert, F., Chen, T. N., and Mandart, E. (1982). Nucleotide sequence of a cloned woodchuck hepatitis virus genome: comparison with the hepatitis B virus sequence. J. Virol. 41:51-65.
6. Summers, J. (1981). The recently described animal virus models for human hepatitis B virus. Hepatology 1: 179-183.
7. Beasley, R. P., Lin, C-C., Hwang, L-Y, and Chien, C-S. (1981). Hepatocellular carcinoma and hepatitis B virus. A prospective study of 22,707 men in Taiwan. Lancet 2:1129-1132.
8. Brechot, C., Pourcel, C., Louise, A., Rain, B., and Tiollais, P. (1980). Presence of integrated hepatitis B virus DNA in cellular DNA of hepatocellular carcinoma. Nature 286:533-535.
9. Shafritz, D. A., Shouval, D., Sherman, H. I., Hadziyannis, S. J., and Kew, M. (1981). Integration of hepatitis B viral DNA into the genome of liver cells in chronic liver disease and hepatocellular carcinoma. N. Engl. J. Med. 305:1067-1073.
10. Ogston, C. W., Jonak, G. J., Rogler, C. E., Astrin, S. M., and Summers, J. (1982). Cloning and structural analysis of integrated woodchuck hepatitis virus sequences from hepatocellular carcinomas of woodchucks. Cell 29:385-394.
11. Rogler, C. E. and Summers, J. Novel forms of woodchuck hepatitis virus DNA isolated from chronically infected woodchuck liver nuclei. J. Virol. (in press).
12. Brechot, C., Hadchouel, M., Scotto, J., Fonck, M., Potet, F., Vyas, G., and Tiollais, P. (1981). State of hepatitis B virus DNA in hepatocytes of patients with HBsAg positive and HBsAg negative liver disease. Proc. Natl. Acad. Sci. (USA) 78:3906-3910.

13. London, W. T. (1981). Primary hepatocellular carcinoma--etiology, pathogenesis, and prevention. Human Path. 12:1085-1097.

MEDICAL CONSEQUENCES OF THE CARRIER STATE

W. Thomas London

Institute for Cancer Research
Fox Chase Cancer Center
Philadelphia, Pennsylvania 19111

Chronic infection with hepatitis B virus (HBV) may lead to the development of chronic hepatitis (both chronic persistent and chronic active), post-necrotic cirrhosis and primary hepatocellular carcinoma (PHC). This statement, which does not sound particularly revolutionary now, is directly contrary to the consensus of experts circa 1960 (1). At that time, it was thought that neither infectious nor serum hepatitis (the terms used then) led to chronic liver disease. Chronic hepatitis and cirrhosis, it was thought, were caused by autoimmune or toxic reactions to unidentified agents.

In the light of present knowledge it is interesting to examine the scientific basis for these conclusions. In the late 1940's and 1950's, when percutaneous liver biopsies came into common use, several investigators performed biopsies on patients who had had episodes of acute hepatitis 5 to 15 years previously (2,3). The usual finding was a histologically normal liver. We now know that hepatitis A infections never result in chronic liver disease, and acute symptomatic hepatitis B infections are usually transient and rarely lead to chronic liver disease. That is, the observations made at that time were valid, but the information collected was not sufficient to draw the correct inferences.

As tests for hepatitis B surface antigen (HBsAg) and other markers of HBV infection became available in the late 1960's, new information was collected which quickly changed the interpretation of the role of hepatitis viruses in chronic liver disease. Table 1 shows the results of two studies done in 1969 in which the insensitive immunodiffusion test was used to assay serum samples from patients with biopsy-proven chronic liver disease. In the study

Table 1. Early studies of HBV and chronic liver disease

Year	Authors	Number of Patients (n)	HBsAg(+)	Patients
1969	Gitnick, et al. (4)	31	3	Chronic active liver disease with cirrhosis
1969	Wright et al. (5)	24	6	Chronic active hepatitis

by Gitnick et al. (4) serial serum samples were available and HBsAg was detected in the initial serum samples of 3 of 31 patients with chronic hepatitis and cirrhosis. Wright et al. (5) in the same year found that 6 of 24 (25%) patients with chronic active hepatitis were HBsAg(+). Both groups of investigators showed that HBsAg remained detectable in the serum of some of these patients for several years. It is now accepted that a considerable proportion (25 to 50%) of the cases of chronic hepatitis and cirrhosis in the United States and western Europe are caused by HBV. Precise figures are not available because, interestingly enough, large sequential series of patients with chronic liver disease, tested by modern methods for the detection of HBV related antigens and antibodies, have not been reported. In endemic areas it appears from case-control studies that almost all cases of chronic liver disease are associated with HBV infection. For example, a study by Hann et al. (6) of 50 consecutive biopsy-verified cases of chronic active hepatitis admitted to the Liver Unit of the National University Hospital in Seoul, South Korea showed that 78% were HBsAg(+) and 93% had antibodies to the hepatitis B virus core antigen (anti-HBc). Of 35 consecutive cases with biopsy-proven cirrhosis, 92% were HBsAg(+) and 100% were anti-HBc(+). In the general population of Korea between the ages of 30 and 59, about 5% are HBsAg(+) and 70% anti-HBc(+).

For the purposes of this paper, the terms chronic carrier and persistent infection with HBV will be used interchangeably. There is some confusion concerning this terminology. Some investigators have used chronic carrier to designate only those individuals with persistent infection who do not have clinical or laboratory abnormalities of the liver. Since such a definition excludes patients from the carrier class as they become ill, it is misleading.

Liver biopsies have been done on chronic carriers. Some studies reported only asymptomatic individuals with normal biochemical tests. Others reported asymptomatic persons and stratified them according to normal or abnormal transaminases. In general, liver biopsies of asymptomatic blood donors showed non-specific reactive hepatitis (7,8,9,10). Overall, about 4% of chronic carriers in the United States or western Europe have chronic active hepatitis or cirrhosis when biopsied (Table 2).

The next question is whether these individuals with mild liver disease are at risk of developing more serious liver disease. In order to answer this question I will briefly summarize the data relating persistent infection with HBV to the etiology of primary hepatocellular carcinoma. (The following points are given in greater detail in reference 11.)

1) The international distribution of the prevalences of HBV carriers roughly correlates with the incidence and mortality of primary hepatocellular carcinoma (PHC).

2) Case-control studies of patients with PHC and appropriate controls from the same locality show large differences. Fifty to 80% of PHC cases in endemic areas are HBsAg(+) compared with 5 to 15% of controls. Eighty to 90% are anti-HBc(+) compared with 15 to 35% of controls. In the United States 30 to 50% of cases are anti-HBc(+) compared with 1 to 5% of controls.

3) There is clustering in families of chronic carriers of HBV and cases of PHC (and other chronic liver diseases). Studies in Senegal, West Africa (12) and South Korea (6) show that 40 to 70% of the mothers of cases of PHC are chronic carriers of HBV. This implies that virus was transmitted by carrier mothers to their offspring and that some of these children subsequently developed PHC.

4) Histopathologic studies have shown that HBsAg is usually found in the liver of patients with PHC. Most investigators find that the viral protein is present in the cytoplasm of non-neoplastic cells and rarely in the tumor cells (13).

5) Several groups have used hybridization methods to search for HBV DNA in human liver cancers. HBV DNA has been found integrated into the genome of the tumor cells by most investigators (14,15). The viral DNA sequences are integrated in a clonal pattern in the tumors. That is, discrete bands are present and the same pattern of bands is present in different parts of the same tumor. This observation implies that integration of viral DNA occurred in the first cancer cell or preceded the development of the initial cancer cell.

6) Epidemiologically, the most convincing and most powerful studies of etiology are prospective. Two large studies are in

Table 2. Biopsy studies of chronic carriers

Year	Authors	Number HBsAg(+)	Chronic Persistent Hepatitis or Non-Specific Hepatitis	Cirrhosis and/or Chronic Active Hepatitis
1974	Vittal et al. (8)	30	28	0
1974	Simon, Patel (7)	9	6	2
1980	DeFranchi et al. (9)	114	82	5
1982	Dragosics et al. (10)	242	166	12

progress--one in Taiwan and a smaller one in Japan. Beasley et al. (16,17) are following 22,707 men in Taiwan who are between the ages of 40 and 60 and are members of the Chinese civil service. The men were recruited into the study between December 1975 and June 1978. They were tested for HBV markers at the time of recruitment and classified as HBsAg(+) or HBsAg(-). As of September 1, 1981, 70 cases of PHC had occurred among the 3454 HBsAg(+) individuals and only one among the 19,253 negative persons (Table 3). This gives a relative risk for carriers to develop PHC that is 390 times that of non-carriers. The carriers accounted for 98% (attributable risk) of all the cases of PHC. Beasley estimates that 40% of the carriers will eventually develop PHC and 10 to 15% will die of cirrhosis.

The second study is being carried out in Japan by Sakuma et al. (18) and includes 3030 employees of the Japan National Railway Co. who were between 40 and 55 years of age at recruitment. Ninety-eight percent were men. They were tested for HBV markers and followed for five years (Table 4). Four died of PHC. All four were HBsAg(+) at the beginning of the study. During the five years this study was done, the Japanese Railways employed about 39,000 employees above the age of 40. Seventeen died of PHC. By applying the findings in the subpopulation to the total population, it is highly likely that all or almost all of these liver cancer deaths were due to HBV infection.

7) Another way to evaluate the risk of liver cancer in individuals persistently infected with HBV is to follow patients with chronic liver disease who are or are not chronic carriers of HBV. In a study of this kind by Obata et al. (19), 7 of 30 HBsAg(+) patients with cirrhosis developed PHC over the three- to five-year period following the diagnosis of cirrhosis, compared

Table 3. Prospective study of asymptomatic carriers of HBV and controls in Taiwan. Entry to study 11/3/75 - 6/30/78. Follow-up to 6/30/81 (Beasley, R. P.) (16,17).

HBsAg status on recruitment	Population at Risk	Incidence of PHC[a]	Deaths		
			PHC[a]	Cirrhosis	Other
HBsAg(+)	3454	70	40	17	48
HBsAg(-)	19,253	1	1	2	199

[a]PHC = primary hepatocellular carcinoma.

Table 4. Deaths during a 5-year follow-up of Japanese railway workers studied by Sakuma et al. (18).

Hepatitis Markers	n	Deaths			% of n
		PHC[b]	Other causes	Total deaths	
HBsAg(+)	202	4	3	7	3.47
anti-HBs(+)[a]	502	0	5	5	1.0
Both (-)	2426	0	19	19	0.8

[a]Antibodies to HBsAg.
[b]PHC = primary hepatocellular carcinoma.

with 5 of 85 HBsAg(-) cirrhotic patients. That is, patients with cirrhosis who are chronic carriers of HBV are at about four times the risk of developing PHC as patients with cirrhosis who are not carriers.

In collaboration with C. Y. Kim, our group is following 245 patients with chronic liver disease diagnosed by biopsy before 1978. All have been infected with HBV. During the observation period of 2 to 12 years, 62 patients have died, all from liver related causes. Thirty-two developed PHC (Table 5). Eighteen had an initial diagnosis of chronic active or chronic persistent hepatitis and 13 of cirrhosis. Therefore, most patients who develop PHC probably pass through a stage of chronic inflammatory liver disease which may be clinically silent.

8) It is likely, therefore, that chronic infection with HBV can result in chronic liver disease and PHC. A virus closely related to HBV has been identified by Summers and his colleagues in woodchucks (20). Snyder and Summers (21) have established that woodchucks persistently infected with the woodchuck hepatitis virus (WHV) develop chronic hepatitis and PHC. Animals not chronically infected with WHV do not develop PHC. As in humans, WHV DNA sequences are found integrated in PHC cells and each tumor appears to be clonal with respect to the WHV integrations (22).

Taking these different lines of evidence together, we conclude that persistent infection with HBV can be considered the "cause" of most cases of primary hepatocellular carcinoma in humans. It also appears to be the cause of most of the cases of chronic hepatitis and cirrhosis in the areas of the world where

Table 5. Development of primary hepatocellular carcinoma from chronic liver disease, a prospective study in Korea (Hann et al.) (unpublished data).

Initial diagnosis	n	Developed PHC	% of n
Chronic hepatitis	145	18	12.4
Cirrhosis	67	13	19.4
Acute hepatitis	33	1	3.0
Total	245	32	13.1

hepatitis B infections are endemic, and a significant factor in many of the cases of chronic liver disease in non-endemic areas. If this is the case, prevention of infection with HBV should, in due course, prevent the development of both chronic liver disease and PHC.

SUMMARY

Retrospective and prospective studies of the relationship of persistent infection with hepatitis B virus to primary hepatocellular carcinoma indicate a causal association. HBV DNA is clonally integrated in the genome of most PHCs. The woodchuck hepatitis virus, WHV, a virus closely related to human HBV, causes a similar range of hepatic diseases in woodchucks as HBV does in humans. Therefore, it is likely that prevention of infection with HBV will also prevent the development of chronic liver disease and PHC.

ACKNOWLEDGMENTS

This work was supported by USPHS grants CA-06551, RR-05539 and CA-06927 from the National Institutes of Health and by an appropriation from the Commonwealth of Pennsylvania.

REFERENCES

1. Chalmers, T. C. and Sebestyen, C. S. (1962). Evidence against infectious hepatitis as a cause of cirrhosis. Trans. Amer. Clin. Climat. Assoc. 74:192.

2. Neefe, J. R., Gambescia, J. M., Kurtz, C. H., Smith, H. D., Beebe, G. W., Jablon, S., Reinhold, J. G., and Williams, S. C. (1955). Prevalence and nature of hepatic disturbance following acute viral hepatitis with jaundice. Ann. Intern. Med. 43:1.
3. Nefzger, M. D. and Chalmers, T. C. (1963). The treatment of acute infectious hepatitis. Ten-year follow-up study of the effects of diet and rest. Am. J. Med. 35:299.
4. Gitnick, G. L., Schoenfield, L. J., Sutnick, A. I., London, W. T., Gleich, G. J., Baggenstoss, A. H., Blumberg, B. S., and Summerskill, W. H. (1969). Australia antigen in chronic active liver disease with cirrhosis. Lancet 2:285.
5. Wright, R., McCollum, R. W., and Klatskin, G. (1969). Australia antigen in acute and chronic liver disease. Lancet 2:117.
6. Hann, H. L., Kim, C. Y., London, W. T., Whitford, P., and Blumberg, B. S. (1982). Hepatitis B virus and primary hepatocellular carcinoma: Family studies in Korea. Int. J. Cancer 30:47.
7. Simon, J. B. and Patel, S. (1974). Liver disease in asymptomatic carriers of hepatitis B antigen. Gastroenterology 66:1028.
8. Vittal, S. B., Dourdourekas, D., Shobassy, N., Gerber, M., Teleschi, J., Szanto, P., Stegman, F., and Clowdus, B. F. (1974). Asymptomatic hepatic disease in blood donors with hepatitis B antigenemia. Am. J. Clin. Pathol. 62:649.
9. DeFranchi, R., D'Armino, A., Vecchi, M., Ronchi, G., del Ninno, F., Pariavacini, A., Ferroni, P., and Zanetti, P. (1980). Chronic asymptomatic HBsAg carriers: Histologic abnormalities and diagnostic and prognostic value of serologic markers of the HBV. Gastroenterology 79:521.
10. Dragosics, B., Franci, P., Denk, H. and Wewalka, F. (1982). Clinical and histological long-term study of HBsAg(+) voluntary blood donors in Austria. Hepatology 2:747.
11. London, W. T. (1981). Primary hepatocellular carcinoma. Human Pathol. 12:1085.
12. Larouze, B., London, W. T., Saimot, G., Werner, B. G., Lustbader, E. D., Payet, M., and Blumberg, B. S. (1976). Host responses to hepatitis B infection in patients with primary hepatic cancer and their families. A case/control study in Senegal, West Africa. Lancet 2:534.
13. Nayak, N. C., Dhar, A., Sachdeva, R., Mittal, A., Seth, H. N., Sudarsanam, D., Reddy, B., Waghaolikar, U. L., and Reddy, C.R.R.M. (1977). Association of human hepatocellular carcinoma and cirrhosis with hepatitis virus surface and core antigens in the liver. Int. J. Cancer 20:643.

14. Shafritz, D., Shouval, D., Sherman, H. I., Hadziyannis, S. J., and Kew, M. C. (1981). Integration of hepatitis B virus DNA into the genome of liver cells in chronic liver disease and hepatocellular carcinoma. N. Engl. J. Med. 305:1067.
15. Brechot, C., Pourcel, C., Hadchouel, M., Dejean, A., Louise, A., Scotto, J., and Tiollais, P. (1982). State of hepatitis B virus DNA in liver diseases. Hepatology 2:275.
16. Beasley, R. P., Lin, C.-C., Hwang, L.-Y., and Chien, C.-S. (1981). Hepatocellular carcinoma and hepatitis B virus. A prospective study of 22,707 men in Taiwan. Lancet 2:1129.
17. Beasley, R. P. (1982). Hepatitis B virus as the etiologic agent in hepatocellular carcinoma - Epidemiologic considerations. Hepatology 2:21S.
18. Sakuma, K., Takara, T., Okuda, K., Tsudo, F., and Mayumi, M. (1982). Prognosis of hepatitis B virus surface antigen carriers in relation to routine liver function tests: A prospective study. Gastroenterology 83:144.
19. Obata, H., Hayashi, N., Motoike, Y., Hisamits, T., Okuda, H., Kubayashi, S., and Nishioka, K. (1980). A prospective study on the development of hepatocellular carcinoma from liver cirrhosis with persistent hepatitis B infection. Int. J. Cancer 25:741.
20. Summers, J., Smolec, J. M., and Snyder, R. (1978). A virus similar to human hepatitis B virus associated with hepatitis and hepatoma in woodchucks. Proc. Natl. Acad. Sci. (USA) 75:4533.
21. Snyder, R. L. and Summers, J. (1980). Woodchuck hepatitis virus and hepatocellular carcinoma, _in_ Viruses in Naturally Occurring Cancers, M. Essex, G. Todaro and H. zur Hausen, eds., pp. 447-457, Cold Spring Harbor Conferences on Cell Proliferation, Cold Spring Harbor Laboratory.
22. Ogston, C. W., Jonak, G. J., Rogler, C. E., Astrin, S. M., and Summers, J. (1982). Cloning and structural analysis of integrated woodchuck hepatitis virus sequences from hepatocellular carcinomas of woodchucks. Cell 29:385.

THE EPIDEMIOLOGY OF HEPATITIS B

Jules L. Dienstag

Gastrointestinal Unit (Medical Services)
Massachusetts General Hospital
and
Department of Medicine
Harvard Medical School
Boston, Massachusetts

INTRODUCTION

Hepatitis B virus (HBV) is an infectious agent that is distributed worldwide and which is perpetuated among humans in a large reservoir of chronic carriers. On a global scale, there are approximately 200 million carriers of HBV, i.e. 5% of the earth's population are infected. The prevalence of infection with this virus, however, is not uniform throughout the world. Instead, there is a geographic gradient in the distribution of HBV, lowest in North America and western Europe, where the virus can be detected in 0.1 to 0.5% of the population, and highest in Southeast Asia and sub-Saharan Africa, where the frequency of current infection may approach 5 to 20% of the population. This skewed distribution parallels that of hepatocellular carcinoma and provides strong epidemiologic evidence for an association between chronic HBV infection and this type of malignancy. If all these carriers were ill and temporarily or permanently disabled, the opportunities for spread to others in the community would be few. Unfortunately, a substantial proportion of infected persons are asymptomatic and, via routine interactions with others, serve as insidious, unrecognized sources for the dissemination of infection. Because those areas of the world with the highest prevalences of infection tend also to be the poorest, prospects for prevention with expensive prophylactic interventions are limited. Moreover, because of increasing travel, primarily by emigration, by persons from high-risk areas to countries with low prevalences of infection (and immunity), the potential exists for introduction of new foci of infection into low-risk areas.

MODES OF TRANSMISSION

A misconception about hepatitis B, arising prior to the availability of HBV serologic markers and reflected in the misnomer "serum hepatitis," was the notion that transmission of this virus required percutaneous exposure to blood, blood products, or virus-contaminated instruments. To be sure, even contemporary re-evaluation with sophisticated seroepidemiologic techniques has confirmed the concept that HBV infection is not transmitted readily from person to person. Furthermore, percutaneous inoculation is one of the most efficient mechanisms implicated in the transmission of HBV infection. The spread of infection with this virus, however, is effected not only by percutaneous but also by apparently nonpercutaneous routes as well.

Percutaneous transmission accounts for approximately half of hepatitis B cases in nonendemic areas (1). Among the sources of such transmission are blood and blood products and contaminated needles and other instruments. Infection may be spread by needlestick, penetration of the integument with any instrument, mucous membrane splash, scarification, tattooing, ear-piercing, etc. There is even some evidence that blood-sucking arthropods such as mosquitos and bedbugs can harbor HBsAg (2,3). Theoretically, they could serve as vectors for transmission of HBV infection, especially in tropical areas in which HBV infection is endemic; however, such transmission has never been documented in natural or experimental settings (4).

The other roughly 50% of hepatitis B cases occur in the absence of an identifiable penetration of the integument or any other exposure to blood or instruments. These cases have been attributed to nonpercutaneous or, alternatively, to covert percutaneous transmission. Although the mechanism(s) of spread under these circumstances have not been defined adequately, HBV in body fluids and secretions may play a role. In patients with HBV infection, hepatitis B surface antigen (HBsAg) can be detected in almost every body fluid including saliva, tears, seminal fluid, vaginal secretions, cutaneous exudates, ascites, pleural fluid, synovial fluid, gastric juice, cerebrospinal fluid, breast milk, urine, and, rarely, feces. Although there is abundant evidence against the infectivity of feces, at least some of these body fluids have been shown to be infectious when administered percutaneously or nonpercutaneously to susceptible experimental animals. Both saliva and semen from HBsAg-positive patients have been used successfully to transmit HBV infection by subcutaneous or intravenous injection to chimpanzees or gibbons; whereas saliva was not infectious when administered via aerosol to nasal or oral surfaces, semen administered intra-vaginally followed by vaginal manipulation to simulate sexual intercourse did lead to infection (Table 1) (5,6,7). Although body fluids are recognized to be potentially infectious, their infectivity is several orders of magnitude lower than the infectivity of blood or serum from the same patient. Nevertheless, when blood from an HBsAg-positive person contains 10^{12} to 10^{13} HBsAg particles per ml of blood and on the order of 10 to 10^3 fewer intact virions, even a several-log-fold diminution in infectious virions will leave body fluids with more than the calculated threshold of 10^6 HBsAg particles sufficient for infectivity (8).

Among the apparently nonpercutaneous modes of HBV transmission, oral ingestion has been documented as a potential route of exposure (9) but one whose relative efficiency is quite low; when successful, oral inoculation is followed by a prolonged incubation period and reduced severity of illness (10). Moreover, evidence from studies in experimental animals suggests that oral transmission of HBV infection requires or is markedly facilitated by an interruption in the integrity of the oropharyngeal mucosa (11). Thus, oral transmission appears to represent covert percutaneous rather than true nonpercutaneous inoculation. The relative inefficiency of oral transmission, the absence of identified cases of food-borne type B hepatitis, and the probable inactivation of HBV by gastro-duodenal secretions suggest that oral transmission is not important in the spread of HBV infection.

Among the various mechanisms for transmission of HBV, the two apparently nonpercutaneous modes considered to have the greatest impact are intimate contact (primarily sexual intimacy) and perinatal spread from an HBsAg-positive mother to her offspring.

Table 1. Infectivity of HBsAg-positive body fluids in experimental animals[a]

Body fluid	Mode of inoculation	Infectious
Saliva	percutaneous	yes
	aerosol	no
Semen	percutaneous	yes
	vaginal	yes

[a]In studies conducted by Alter and Bancroft and colleagues (5,6,7).

Sexual intimacy with many different partners appears to be associated with a very high risk of exposure, as seen in promiscuous homosexual men (12). Although such exposure does not involve readily identifiable percutaneous exposure to blood, the opportunities for exchange of secretions are many between sexual partners. Moreover, minute bleeding resulting from trauma to the rectal mucosa has been detected in almost a third of homosexual men evaluated in clinics for sexually transmitted diseases (13). Thus, the likelihood is high that this apparently nonpercutaneous mode of HBV transmission relies on covert percutaneous inoculations. The same may apply to perinatal spread. Although the mechanism for perinatal transmission is unknown, most cases occur after an incubation period consistent with acquisition during labor and delivery. Potentially, HBV infection may result from placental breaks which permit mixing of the maternal and fetal circulation; from ingestion of amniotic fluid; from skin abrasions, mucous membrane penetration, or ingestion of maternal blood during passage through the birth canal or even from the exposure to maternal blood that occurs during delivery by Caesarian section, to cite several suggested sources (14). In short, these nonpercutaneous routes probably rely on obscure percutaneous penetrations.

These two nonpercutaneous modes--sexual intimacy and perinatal transfer--plus percutaneous transmission constitute the three most important routes for the spread of HBV infection. The relative importance of these three mechanisms, however, varies as a function of geographic locale, prevalence of HBsAg in the population, and occupational or recreational factors. For example, percutaneous transmission is the most important mode of spread in areas of low HBV endemicity among health care personnel and among those who inject themselves with illicit drugs. The other identifiable route of transmission in low-endemicity areas, such as North America and western Europe, is sexual contact, as reflected by frequent HBV infections among promiscuous homosexual males and between spouses when one partner is infected. Still other mechanisms of spread must be operative in areas of low HBsAg prevalence; as many as 40% of sporadic acute hepatitis B cases identified in a study by the Centers for Disease Control occurred in persons with no demonstrable predisposing risk factor or exposure (11). Most HBV infections in such nonendemic countries occur during adolescence and early adulthood, when opportunities for percutaneous, sexual and occupational exposures first materialize. In marked contrast, in developing countries and other locales of high hepatitis B endemicity, e.g. the Far East and sub-Saharan Africa, perinatal transmission is the most common mode of HBV spread and, because of the high frequency of HBV infection in these parts of the world, the mode of HBV transmission contributing most to its perpetuation in humans. Despite this large contribution of perinatal spread, other mechanisms in these endemic areas play a role in disseminating HBV infection beyond infancy. In parts of Africa, for example, the frequency of HBV infection increases dramatically when children reach the toddler stage. Circumstantial evidence suggests that intimate contact among toddlers, perhaps via transfer of contaminated serous fluid from oozing cutaneous rashes and abrasions, and/or spread of infection among siblings is instrumental in maintaining the high frequency of hepatitis B surface antigenemia in the population (11,15,16). In addition to such "horizontal" spread, percutaneous exposures also appear to occur in certain high-endemicity areas. In Southeast Asia, injections with contaminated needles in pre-school children have been implicated in the amplification of HBV exposure (17). As predicted from these modes of spread, most HBV infections in endemic areas occur during infancy and early childhood. In areas of moderate endemicity, presumably, HBV infections occur primarily during childhood rather than at birth and primarily by horizontal means; however, these areas of the world have not been as adequately characterized seroepidemiologically. Even in areas of low endemicity, horizontal intrafamily spread occurs (18) but is not common. These geographic differences in the mechanisms of HBV transmission are summarized in Table 2.

Finally, environmental contamination with HBV-infected blood or body fluids has been suspected as a link in the chain of HBV transmission. Both antigenicity and infectivity of HBV are retained when blood containing the virus is allowed to dry on environmental surfaces (19,20). Hepatitis B surface antigen has been recovered in the absence of visible blood from environmental surfaces in hemodialysis units and clinical laboratories, including laboratory marking pens and laboratory coats (21,22). Such environmental contamination has been postulated to account for

Table 2. Geographic variation in mechanisms of HBV transmission

Endemicity	Location	Age at time of infection	Mode of transmission
Low	N. America W. Europe	early adulthood	sexual percutaneous other
Moderate	Mediterranean E. Europe	childhood	horizontal
High	Africa Far East	birth toddler stage pre-school	perinatal horizontal

the frequency with which health care workers harbor antibody to HBsAg in the absence of antibody to hepatitis B core antigen, i.e. a serologic pattern more consistent with immunization by HBsAg--presumably from HBsAg on environmental surfaces--than with infection by HBV (23). Aerosolization does not contribute to transmission of HBV infection (24).

EFFICIENCY OF TRANSMISSION

Efficiency of HBV transmission varies considerably as a function of the following factors: virus concentration, volume of the inoculum, duration of exposure, route of inoculation and type of exposure, and susceptibility of contacts.

The likelihood of HBV infection after exposure is dependent on the concentration of infectious virions in the inoculum. Among the indicators of virus concentration are the presence and number of detectable virions, DNA polymerase activity, HBV DNA, titer of HBsAg, and the presence of hepatitis B e antigen (HBeAg). Practically, the best and simplest indicator of relative infectivity is HBeAg. Among those sustaining an HBsAg-positive needlestick or comparable percutaneous/transmucosal penetration, the likelihood of becoming infected with HBV is 10 to 20% if the inoculum contains HBeAg but only 1 to 2.5% if the inoculum lacks HBeAg (25, 26). Similarly, the frequency of HBV infection approaches 90 to 100% among babies born to HBsAg-positive mothers who are also HBeAg-positive but is no greater than 10 to 15% among babies born to HBsAg-positive but HBeAg-negative mothers (27,28,29). The relation between HBeAg and infectivity of HBV was demonstrated in an experiment conducted by Shikata et al. (30) who made serial dilutions of HBeAg-positive and HBeAg-negative blood containing HBV. When inoculated into susceptible chimpanzees, HBeAg-positive blood remained infectious, albeit with long incubation periods for very dilute inocula, even when diluted 100 million fold. In contrast, blood containing anti-HBe was barely infectious; chimpanzees inoculated with undiluted serum did become infected after a very long incubation period but diluted serum was no longer infectious (Table 3).

The role of inoculum size is reflected in the difference between the frequencies of HBV infection in persons exposed by being transfused with an entire unit of HBsAg-positive blood (31) and in persons exposed with a minute drop of HBsAg-positive blood on a contaminated needle (25,32,33). After a large-volume inoculum such as a transfusion, the likelihood of HBV infection is as high as 75%; after a needlestick, a very small-volume inoculum, the likelihood of successful infection is less than 15% (Table 4). Similarly, the higher the infecting dose, the shorter the incubation period (34) and, potentially, although less well documented,

Table 3. Relation of HBeAg to infectivity of hepatitis B virus[a]

HBsAg(+) serum inoculum	Outcome in inoculated chimpanzees[b] (Weeks after inoculation)	
	HBsAg(+)	ALT elevated
HBeAg(+)		
undiluted	+ (1)	-
10^{-1}	+ (4-7)	+ (7-9)
10^{-4}	+ (8-25)	+ (12-24)
10^{-8}	+ (13-29)	+ (18-26)
anti-HBs(+)		
undiluted	+ (19-20)	-
10^{-2}	-	-
10^{-5}	-	-

[a]From Shikata et al. (30).
[b]+ = successful infection; - = uninfected; ALT = alanine aminotransferase.

the greater the severity of acute illness. As an illustration, when the frequency of transfusion-associated hepatitis type B was reduced by the introduction of HBsAg screening in blood banks, the fatality rate for acute hepatitis B dropped from 1% to a level on the order of 0.1%. This change appears to reflect the marked decrease in dose of virus inocula when transfusion became a less common mode of HBV transmission.

Although percutaneous inoculation is the most efficient mode of transmission, the 10 to 15% frequency of HBV infection after penetration of the skin with an HBsAg-positive needle or other instrument is surprisingly low. Even when the HBsAg source is HBeAg-positive, the frequency of infection in the inoculee is no greater than 10 to 20%, as stated above. This lower-than-expected frequency is a reflection not only of the lack of true penetration among those reporting skin puncture, but also of the relative inefficiency of an exposure occurring during a brief instant in time. In contrast, sexual exposure to a spouse with acute HBV infection, although probably less efficient than direct percutaneous exposure and likely to be followed by a longer incubation period, is more likely than needlestick to result in HBV infection. Among spouse contacts of patients with acute hepatitis B, 23 to 42% have been reported to acquire HBV infection (including symptomatic and subclinical cases) from their sexual partners (35,36) (Table 4). The high likelihood of transmission in this

Table 4. Relative efficacy of HBV infection after different modes of exposure

Mode of exposure	Frequency of all HBV events[a]
Transfusion (31)	75%
Needlestick (25,32,33)	9-15%
Sexual (35,36)	23-42%

[a]Including clinical and subclinical infections. For needlestick, the frequencies of HBV infection listed include those recorded with or without post-exposure globulin prophylaxis. For sexual exposure, data are derived from observations in spouses of patients with acute hepatitis B who received no prophylaxis (36) or who received standard immune globulin (35).

setting appears to result not necessarily from the efficiency of any single, instantaneous contact but from the cumulative frequency and multiplicity of contacts likely to characterize a sexually intimate relationship.

The route of inoculation can play a role in determining the efficiency of transmission, reflected in the length of the incubation period (10). For example, in one report (37), after percutaneous inoculation, detectable HBsAg appeared in the circulation as early as one week after inoculation, and serum alanine aminotransferase (ALT) became elevated as early as six weeks after exposure. In contrast, after oral inoculation, the incubation period was much longer; HBsAg was detected two months after inoculation, and elevated ALT activity occurred three months after inoculation. This difference in efficiency between percutaneous and oral inoculation may represent differences in the dose of virus reaching the circulation and the liver.

Of course, efficiency of transmission relies as well on the susceptibility of exposed persons. Although the promiscuous lifestyle of male homosexuals predisposes them to high HBV attack rates, their high prevalence of current and previous HBV infection, approximating 70%, precludes efficient transmission of infection to the group as a whole. If 100 homosexual men were exposed to an HBsAg carrier, even if the efficiency of infection were as high as 40% (and it may not be if there is only a single encounter), only 30% of the cohort would be susceptible, and only 12% would become infected.

POPULATION GROUPS WITH ENHANCED RISK OF HBV EXPOSURE

Population groups with enhanced risk of exposure to and infection with HBV include clients and staff of custodial institutions for the mentally disabled; inhabitants of other closed institutions such as prisons, where intravenous drug use and homosexuality contribute to the risk; staff and patients of hemodialysis units; other health care personnel exposed frequently to patients' blood and blood derivatives; recipients of frequent or multiple transfusions, including hemophiliacs, thalassemics, and immunosuppressed patients requiring blood component therapy; sexually promiscuous persons, most notably promiscuous homosexual men; users of illicit intravenous drugs; sexual and household contacts of persons with chronic HBV infection; sexual partners of patients with acute type B hepatitis; newborns of mothers who are HBsAg-positive during late pregnancy; and persons from areas in which the prevalence of HBV infection is high, including persons in endemic areas, refugees from these endemic areas, and lower socioeconomic populations living under crowded, substandard conditions in otherwise low-prevalence countries.

Certain subgroups of the population are more likely not only to become infected, but also to remain chronically infected after acute infection. These include persons with Down's syndrome, patients with immunologic (primarily cellular) deficiencies, chronically hemodialyzed patients, and those infected during infancy and early childhood, especially in endemic areas. The fact that, given similar rates of exposure to HBV, males are more likely than females to remain chronically infected suggests that male sex favors HBsAg carriage. Similarly, the high frequency of chronic HBV infection in Southeast Asia and parts of Africa suggests that geographic or racial factors may predispose to chronic HBV infection.

CONCLUSION

Our understanding of the mechanisms of HBV transmission has evolved and matured and is well beyond the simplistic notion, codified by the designation "serum hepatitis," that percutaneous/transmucosal penetration with blood or blood-contaminated instruments is required. Both percutaneous and apparently nonpercutaneous modes of HBV transmission occur; their relative importance varies among geographic locales, and cultural, environmental, occupational, and recreational factors determine the peak age of exposure and infection. Active immunization with a widely available, readily affordable hepatitis B vaccine is a high priority for these populations at risk. Appropriate timing of vaccination, prior to or no later than the period of peak risk, is crucial.

REFERENCES

1. Prince, A. M., Hargrove, R. L., Szmuness, W., Cherubin, C. E., Fontana, V. J., and Jeffries, G. H. (1970). N. Engl. J. Med. 282:987.
2. Metselaar, D., Blumberg, B. S., Millman, I., Parker, A. M., and Bagshawe, A. F. (1973). Lancet 2:758.
3. Wills, W., Larouze, B., London, W. T., Millman, I., Werner, B. G., Ogston, W., Pourtaghva, M., Diallo, S., and Blumberg, B. S. (1977). Lancet 2:217.
4. Berquist, K. R., Maynard, J. E., Francy, D. B., Sheller, M. J., and Schable, C. A. (1976). Am. J. Trop. Med. Hyg. 25:730.
5. Alter, H. J., Purcell, R. H., Gerin, J. L., London, W. T., Kaplan, P. M., McAuliffe, V. J., Wagner, J., and Holland, P. V. (1977). Infect. Immun. 16:928.
6. Bancroft, W. H., Snitbahn, R., Scott, R. M., Tingpalapong, M., Watson, W. T., Tanticharoenyos, B., Karwacki, J. J., and Srimarut, S. (1977). J. Infect. Dis. 135:79.
7. Scott, R. M., Snitbahn, R., Alter, H. J., and Bancroft, W. H. (1980). J. Infect. Dis. 142:67.
8. Gerety, R. J., Tabor, E., Hoofnagle, J. H., Mitchell, F. D., and Barker, L. F. (1978). in Viral Hepatitis, G. N. Vyas, S. N. Cohen and R. Schmid, eds., Franklin Institute Press, Philadelphia, p. 121.
9. Krugman, S., Giles, J. P., and Hammond, J. (1967). JAMA 200:365.
10. Krugman, S. and Giles, J. P. (1970). JAMA 212:1019.
11. Francis, D. P., Favero, M. S., and Maynard, J. E. (1981). Semin. Liver Dis. 1:27.
12. Szmuness, W., Much, M. I., Prince, A. M., Hoofnagle, J. H., Cherubin, C. E., Harley, E. J., and Block, G. H. (1975). Ann. Intern. Med. 83:489.
13. Reiner, N. E., Judson, F. N., Bond, W. W., Francis, D. P., and Petersen, N. J. (1982). Ann. Intern. Med. 96:170.
14. Lee, A. K. Y., Ip, H. M. H., and Wong, V. C. W. (1978). J. Infect. Dis. 138:668.
15. Whittle, H. C., Bradley, A. K., McLauchlan, K., Ajdukiewicz, A. B., Howard, C. R., Zuckerman, A. J., and McGregor, I. A. (1983). Lancet 1:1203.
16. Prince, A. M., White, T., Pollock, N., Riddle, J., Brotman, B., and Richardson, L. (1981). Infect. Immun. 32:675.
17. Beasley, R. P., Hwang, L-Y., Lin, C-C., Leu, M-L, Stevens, C. E., Szmuness, W., and Chen, K-P. (1982). J. Infect. Dis. 146:198.
18. Leichtner, A. M., LeClair, J., Goldmann, D. A., Schumacher, R. T., Gewolb, I. H., and Katz, A. J. (1981). Ann. Intern. Med. 94:346.

19. Bond, W. W., Petersen, N. J., and Favero, M. S. (1977). Health Lab. Sci. 14:235.
20. Bond, W. W., Favero, M. S., Petersen, N. J., Gravelle, C. R., Ebert, J. W., and Maynard, J. E. (1981). Lancet 1: 550.
21. Dankert, J., Uitentuis, J., Houwen, B., Tegzess, A. M., and van der Hem, G. K. (1976). J. Infect. Dis. 134:123.
22. Lauer, J. L., VanDrunen, N. A., Washburn, J. W., and Balfour, H. H., Jr. (1979). J. Infect. Dis. 140:513.
23. Dienstag, J. L. and Ryan, D. M. (1982). Am. J. Epidemiol. 115:26.
24. Petersen, N. J., Bond, W. W., Marshall, J. H., Favero, M. S., and Raij, L. (1976). Health Lab. Sci. 13:233.
25. Seeff, L. B., Wright, E. C., Zimmerman, H. J., Alter, H. J., Dietz, A. A., Felsher, B. F., Finkelstein, J. D., Garcia-Pont, P., Gerin, J. L., Greenlee, H. B., Hamilton, J., Holland, P. V., Kaplan, P. M., Kiernan, T., Koff, R. S., Leevy, C. M., McAuliffe, V. J., Nath, N., Purcell, R. H., Schiff, E. R., Schwartz, C. C., Tamburro, C. H., Vlachevic, Z., Zemel, R., and Zimmon, D. S. (1978). Ann. Intern. Med. 88:285.
26. Werner, B. G. and Grady, G. F. (1982). Ann. Intern. Med. 97:367.
27. Okada, K., Kamiyama, I., Inomata, M., Imai, M., Miyakawa, Y., and Mayumi, M. (1976). N. Engl. J. Med. 294:746.
28. Stevens, C. E., Neurath, R. A., Beasley, R. P., and Szmuness, W. (1979). J. Med. Virol. 3:237.
29. Shiraki, K., Yoshihara, N., Sakurai, M., and Kawana, T. (1980). J. Pediat. 97:768.
30. Shikata, T., Karasawa, T., Abe, K., Uzawa, T., Suzuki, H., Oda, T., Imai, M., Mayumi, M., and Moritsugu, Y. (1977). J. Infect. Dis. 136:571.
31. Gocke, D. J. (1972). JAMA 219:1165.
32. Grady, G. F., Lee, V. A., Prince, A. M., Gitnick, G. L., Fawaz, K. A., Vyas, G. N., Levitt, M. D., Senior, J. R., Galambos, J. T., Bynum, T. E., Singleton, J. T., Clowdus, B. F., Akdamar, K., Aach, R. D., Winkelman, E. I., Schiff G. M., and Hersh, T. (1978). J. Infect. Dis. 138:625.
33. Szmuness, W. (1979). J. Med. Virol. 4:327.
34. Barker, L. F. and Murray, R. (1972). Am. J. Med. Sci. 263:27.
35. Redeker, A. G., Mosley, J. W., Gocke, D. J., McKee, A. P., and Pollack, W. (1975). N. Engl. J. Med. 293:1055.
36. Koff, R. S., Slavin, M. M., Connelly, L. J. D., and Rosen, D. R. (1977). Gastroenterol. 72:297.
37. Krugman, S. (1979). N. Engl. J. Med. 300:625. (Letter).

THE MORPHOLOGIC EXPRESSION OF HEPATITIS B

Evgenya Skarinsky and Emanuel Rubin

Department of Pathology and Laboratory Medicine
Hahnemann University School of Medicine
Philadelphia, Pennsylvania 19102

There are at least three different viruses responsible for the common varieties of hepatitis: the hepatitis A virus (HAV), the hepatitis B virus (HBV), and an unidentified virus or group of viruses called non-A, non-B (NANB). The symptoms and extent of liver injury produced in acute viral hepatitis vary from a mild, non-icteric, subclinical illness to acute icteric hepatitis, massive necrosis and hepatic failure. It is estimated that 5-10% of adults infected with HBV develop a chronic carrier state, manifested by the presence of the surface antigen (HBsAg) in the blood. The severity and variability of clinical and pathological forms of HB are related to the intensity and virulence of the infection and the quality and quantity of the immune response. Although a role for immune responses in the pathogenesis of hepatic injury is not conclusively proved, the data, such as the consistency of humoral and cellular responses, the presence of abnormal plasma immunoregulatory proteins (rosette inhibitory factor, RIF) (15,16,17), decreased suppressor cell activity (21), and the presence within the liver itself of potent immunosuppressive molecules, are compatible with explanations for recovery, acute and chronic injury, and the carrier state. Following inoculation or ingestion, the virus reaches the liver, where extensive virus replication occurs within the hepatocytes. Hepatitis B virus seems to exert no direct cytopathic effect on the infected liver cells. Instead, in some way it may alter the structure of the membranes of infected cells, thus evoking an immune reaction to these altered cells. If the immune reaction is adequate, anti-HBV antibody leads to clearance of the virus from the blood and the liver. The removal of the virus allows normal regulation of cellular immunity, and the suppressor cells may prevent a cellular immune reaction, leading to the recovery. Massive necrosis occurs, presumably, when the immune response is overwhelming. On the other hand, the majority of chronic HBV carriers have a only a mild subclinical hepatitis. This observation suggests that patients with chronic HBV do not develop adequate antibody responses to HB viral antigen, and thus do not clear the virus (75). The persistence of HBV genome may result in impaired function of suppressor cells, with progressive cellular attack on hepatocyte antigens and perhaps viral antigens. The end result would be chronic active hepatitis, smoldering or recurrent episodes of injury that may eventually result in cirrhosis. In the asymptomatic HBV carrier there is also an impaired antibody response to the virus, but the normal function of suppressor cells is preserved, leading to suppression of antiviral and antihepatocyte attack systems.

ACUTE VIRAL HEPATITIS

Approximately 30-40% of adults infected with HBV develop clinically apparent acute hepatitis. In most cases after a long incubation period (60-180 days) the disease is manifested by the prodromal period, characterized by nonspecific symptoms, such as malaise, fatigue, anorexia, and distaste for cigarettes. In 10-20% of patients manifestation of a "serum sickness"-like syndrome is observed (26). These symptoms have been ascribed to circulating immune complexes of viral antigen and antibody. This pre-icteric phase is marked by rapid rise of serum transaminases. In the icteric phase of hepatitis, the symptoms usually diminish, and transaminases begin to decline as the serum bilirubin level rises. The majority of people (50-60%) exposed to the virus never develop jaundice and symptoms of hepatitis. Anicteric mild infection occurs more frequently in children (38).

The pathologic appearance of the liver in viral hepatitis is characterized by a combination of liver cell injury, inflammatory reaction and hepatocellular regeneration, all of which seem to be present simultaneously. The earliest liver cell injury is reflected in diffuse swelling of the hepatocytes throughout the lobule. When cellular enlargement is extreme, such cells contain only thin strands of eosinophilic cytoplasm; the cells look empty and are called "balloon" cells (Figure 1). The enlargement of these cells is mainly the result of edema and marked dilation of the cisternae of the endoplasmic reticulum, with detachment of polysomes and a decrease in the number of ribosomes. It is also in part ascribed to the loss of glycogen from the cytoplasm (52,62). When nuclear lysis in "balloon" cells is complete, hepatocytolysis occurs, leading to the disappearance of liver cells, so-called drop-out necrosis.

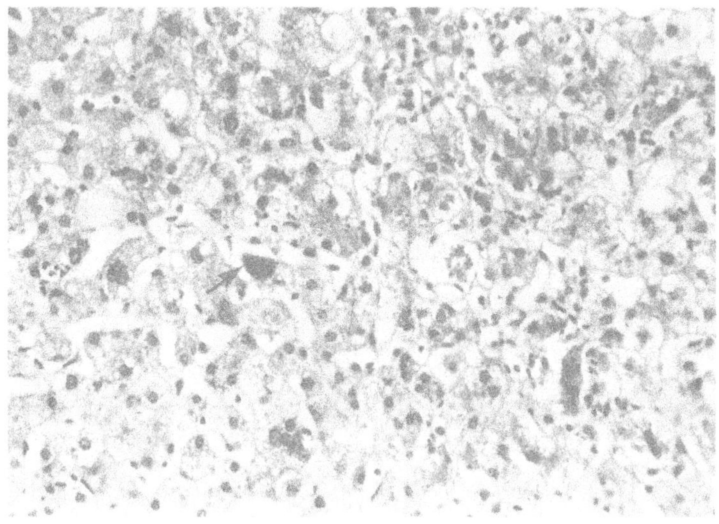

Figure 1. Acute viral hepatitis. Diffuse swelling of hepatocytes (ballooning degeneration) is noted throughout the lobule. A monocytic infiltrate and a single acidophilic body (arrow) are present. (Hematoxylin and eosin stain. x 206.)

Another characteristic type of necrosis which occurs in viral hepatitis is "acidophilic degeneration." The cytoplasm of some hepatocytes undergoes contraction and coagulation, and the nuclei become pyknotic and disappear. Shrunken cells are extruded into the sinusoids as dense, round eosinophilic bodies (Figure 2). These "acidophilic" or Councilman-like bodies appear to contain the remnants of cell organelles by electron microscopy.

Acute liver cell damage occurs nearly simultaneously with an inflammatory reaction, which is recognized morphologically by reticuloendothelial and lymphocytic responses. The cells of the mononuclear phagocytic system are the chief scavengers of hepatic injury. A highly characteristic feature of acute viral hepatitis is hypertrophy and proliferation of Kupffer cells and portal macrophages. Following the destruction of hepatocytes, these cells become loaded with stainable iron, lipofuscin and ceroid pigment, derived from dead liver cells. Lymphocytes and other mononuclear cells accumulate within the focal areas of cell necrosis, and infiltrate between surviving hepatocytes and adjacent liver cell plates. During the course of acute viral hepatitis, most or all liver cells are thought to be affected. The injury varies in intensity and is most severe in the centrilobular areas, thus giving an appearance of focal or spotty distribution of cell

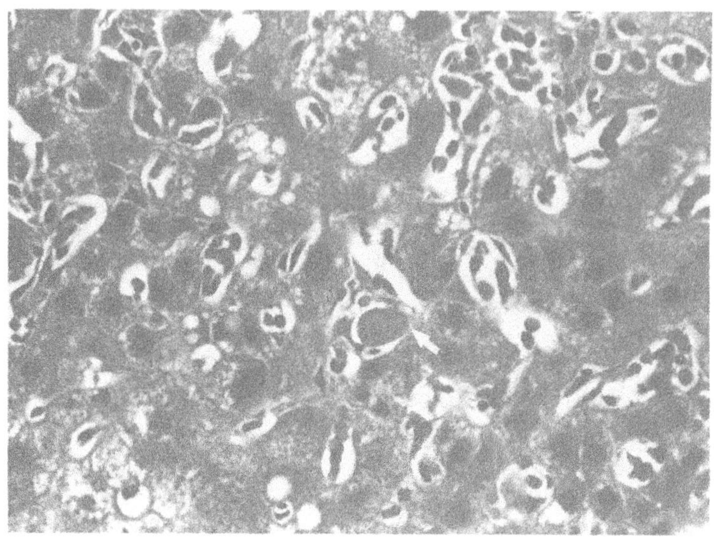

Figure 2. Acute viral hepatitis. Some hepatocytes are shrunken and deeply stained (arrowheads); these are precursors of acidophilic bodies. A single rounded acidophilic body (arrow) is seen. Hypertrophy and proliferation of Kupffer cells are present. (Hematoxylin and eosin stain. x 450.)

changes (Figure 3). These lobular changes and the distribution of liver cell drop-out (hepatocytolysis) are the most characteristic features of acute viral hepatitis.

Universal triaditis, with a tendency for the inflammatory infiltrate to extend into adjacent parenchyma, is also characteristic, requiring differentiation from chronic active hepatitis (Figure 4). Careful evaluation of the parenchymal changes and clinical history help to distinguish these entities.

Regeneration is represented by the presence of rare mitotic figures, enlargement of nuclei, nucleolar prominence, and bi- and tri-nucleated liver cells. The regenerative changes are more prominent in the periportal hepatocytes (48). Because degeneration and regeneration in the liver occur concomitantly, the relationship of the portal tracts and central veins remains preserved. The disarray of cell plates typically occurs within the boundaries of the individual lobules (48). Cholestasis is usually mild and is most prominent in the center of the lobules (zone 3 of Rappaport) (58).

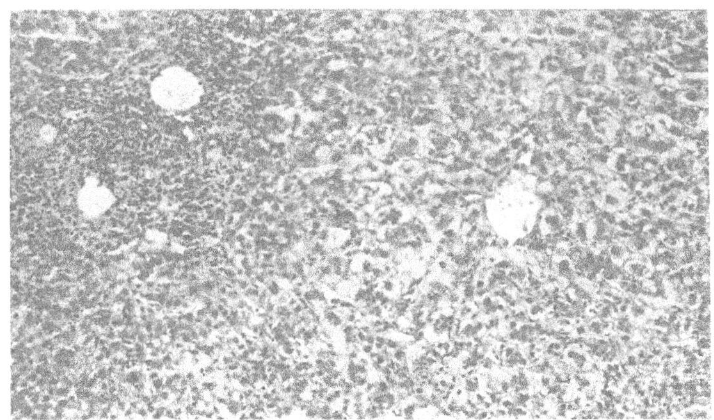

Figure 3. Acute viral hepatitis. Ballooning degeneration, most severe in centrilobular areas, inflammatory reaction and hepatocellular regeneration are present simultaneously, thus giving an appearance of spotty distribution of cell changes. (Hematoxylin and eosin stain. x 140.)

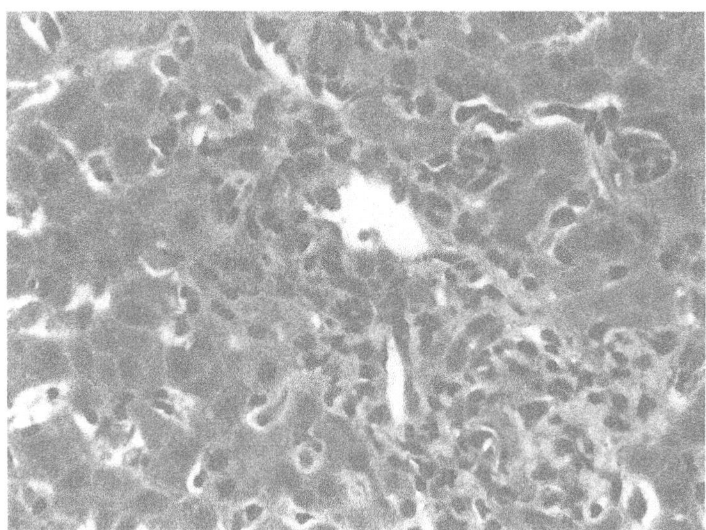

Figure 4. Acute viral hepatitis. A small portal tract is infiltrated by a variety of inflammatory cells, mostly mononuclear. Hypertrophied Kupffer cells and portal macrophages are also present. (Hematoxylin and eosin. x 450.)

Sometimes the clinical picture and laboratory data simulate those associated with mechanical obstruction of the bile ducts. In this cholestatic variant the main histopathologic features are extensive bile retention in the cytoplasm of the hepatocytes and bile plugs in dilated canaliculi (Figure 5). The portal areas are edematous and infiltrated with inflammatory cells, predominantly neutrophilic leukocytes. Ballooning degeneration is usually conspicuous (33).

In most cases of acute HB, the disease subsides within 3-4 weeks. Although there still is minimal focal intralobular inflammation accentuated near the central veins, in this stage of resolving or residual hepatitis, inflammation is mainly confined to the portal tracts, which show well-defined limited plates (20). Degeneration of hepatocytes is minimal or absent, and regeneration is prominent.

Complete clinical recovery occurs in 80% of cases of self-limited infection in 3-4 months and in 90-95% of patients within 6 months (35).

ACUTE VIRAL HEPATITIS WITH BRIDGING NECROSIS

A more severe from of acute viral hepatitis, in which the typical initial symptoms are prolonged and in which the malady becomes progressively debilitating, is variably called "subacute hepatic necrosis," "subacute hepatitis," "submassive necrosis," "impaired regeneration syndrome," or "confluent hepatic necrosis." The names are confusing because some emphasize only the clinical aspects, some only the morphologic presentation, and some only the pathogenesis of the disease. The names "subacute hepatitis" or "subacute hepatic necrosis" are also applied to changes in chronic active hepatitis (9,69), and may have prognostic implications in this condition. Since the definitive diagnostic procedure is a liver biopsy, it seems preferable to use the term "acute viral hepatitis with bridging necrosis."

Significant necrosis is uncommon in typical acute viral hepatitis. When death of groups of adjacent liver cells occurs, the necrosis is called confluent. It may be irregular and may even involve as much as a third of a lobule, but still may be seen in a typical case of acute viral hepatitis. In addition to typical morphologic features described above for classic acute viral hepatitis, the characteristic lesion of this variant of acute hepatitis is intralobular and interlobular zones of necrosis and collapse, bridging adjacent portal triads or central veins, and thus giving an appearance of an irregular branched pattern (Figure 6). Central and mid-zones, (zones 3 and 2 of Rappaport), are commonly involved

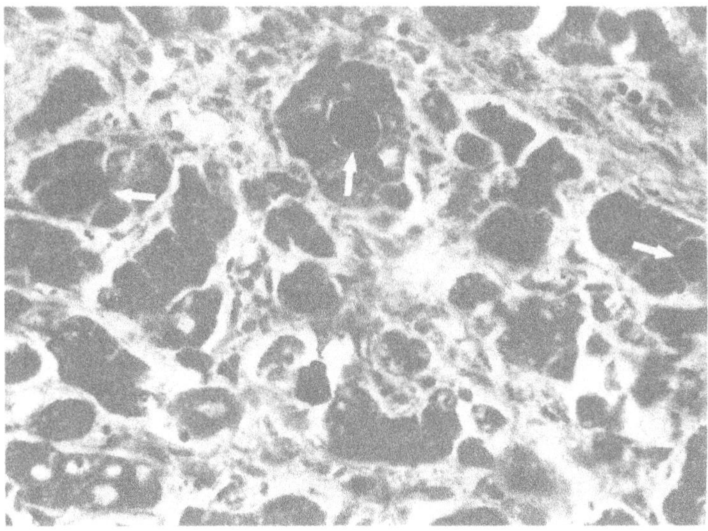

Figure 5. Acute viral hepatitis. Extensive bile stasis in the cytoplasm of the hepatocytes and bile plugs in dilated canaliculi (arrows) are noted. Drop-out necrosis is evident. (Hematoxylin and eosin stain. x 450).

(58). Necrosis of the hepatocytes results in collapse and condensation of the reticulum framework and the appearance of the so-called "passive septa" as opposed to the "active septa" formed by the proliferation of fibroblasts. "Passive" septa are reversible. The formation of "active" septa leads to cirrhosis (54,71,24). Collapse of the reticulum framework is best appreciated in liver

Figure 6. Acute viral hepatitis with severe bridging necrosis. Collapse of the reticulum framework has led to the formation of "passive" septa between central veins and portal tracts. (Hematoxylin and eosin stain. x 50.)

biopsies stained with the silver reticulum stain, while a trichrome stain may be negative or very weakly positive.

Bridging necrosis is a manifestation of severe liver damage and may lead to fatal hepatic failure or to the development of cirrhosis. In one study death occurred in 19% of patients and cirrhosis developed in 37% of the 52 with bridging hepatic necrosis (9). Nevertheless, this lesion is compatible with complete clinical recovery (54% in the same study).

The severity of the disease is clinically represented by the development of the ascites, edema and hepatic coma during the acute stage (the first six months of the disease) and are often the only clinical features diagnostic of this lesion (9). The patients with this form of viral hepatitis are usually over 40 years of age and tend to have an insidious onset of the disease, more deep and prolonged jaundice and more common depression of serum albumin and prothrombin levels (9). It had been suggested that bridging hepatic necrosis reflects the failure of surviving hepatocytes to regenerate (48).

ACUTE VIRAL HEPATITIS WITH MASSIVE NECROSIS

Rarely (less than 1%) patients with acute viral hepatitis develop a fulminant course (59). In such cases, one or two weeks after the onset of symptoms, acute hepatic failure develops. Few patients survive this malady, the mortality being reported as 65-90%. At autopsy the liver may exhibit a normal or reduced weight, depending on the length of survival of the patient. The liver is usually a soft, flabby mass covered by a wrinkled capsule. All zones of the lobule are necrotic, thus justifying the term "massive necrosis." A reticulum framework is the only remnant of the hepatic lobule. Curiously, the inflammatory reaction is generally scanty. Scattered small groups of surviving hepatocytes may be found around the portal tracts. Regenerating cells, frequently arranged in duct-like structures, and proliferated ductules ("necholangioles") (51) are prominent in these areas (Figure 7). In patients who survive fulminant hepatitis, hepatocytes probably start to regenerate from these structures and grow along the preserved but collapsed reticulum fibers. There is no unusual predisposition to the development of the cirrhosis among fulminant hepatitis survivors (14). Complete morphologic and clinical recovery and chronic active hepatitis are both reported (31,36).

CHRONIC VIRAL HEPATITIS

A characteristic of HB is the development of a chronic carrier state, identified by the continued detection of HBsAg in

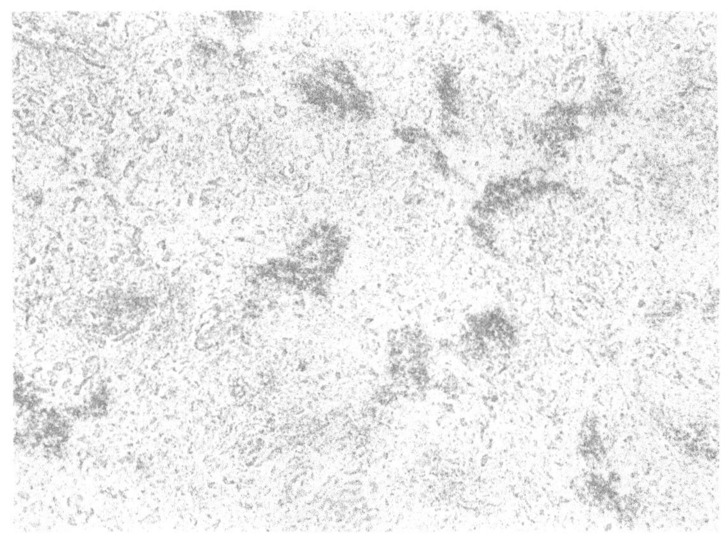

Figure 7. Acute viral hepatitis with massive necrosis. All zones of the lobule are necrotic. Regenerating cells, arranged in duct-like structures, and proliferated bile ductules are prominent. The inflammatory reaction is scanty. (Hematoxylin and eosin stain. x 50.)

the serum, with or without progression to chronic disease. Strictly speaking, the term "carrier" has usually been reserved for individuals who have neither symptoms nor significant liver disease. Because of natural history, clinical course, and serologic manifestations of the chronic HBsAg carrier state, two main types are distinguished. One is the asymptomatic serological carrier, with HBV antigenemia, normal liver function tests (30, 67), and an essentially normal liver biopsy. The second type includes asymptomatic and symptomatic patients with abnormal liver function tests and variable histological alterations. The histopathologic feature common to both types of chronic HBsAg carriers is the presence of "ground glass" hepatocytes (67,43), cells which are larger than normal and have a pale homogeneous cytoplasm and nuclei frequently displaced toward the periphery (Figure 8). The characteristic appearance of the cytoplasm results from proliferated endoplasmic reticulum containing HBsAg (29), in this case 20-30 nm diameter filamentous structures of various lengths (25). HBsAg can also be demonstrated in liver cells by Shikata's orcein staining, immunofluorescence and the peroxidase-antiperoxidase method (25,32,61). The classic asymptomatic HBsAg "carrier," incidentally found to have HBsAg in the serum, usually shows large numbers of ground glass cells and little inflammation in the liver biopsy (Figure 8).

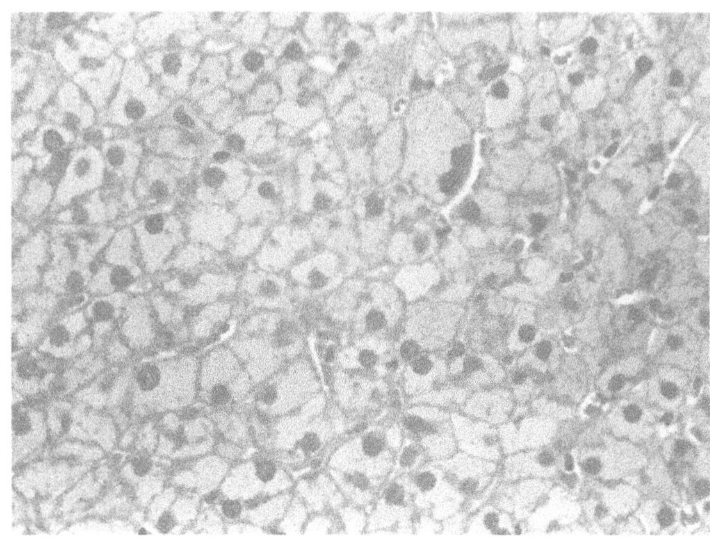

Figure 8. Chronic asymptomatic HB carrier state. Large "ground glass" hepatocytes, with pale homogeneous cytoplasm and nuclei displaced toward the periphery, are present. (Hematoxylin and eosin stain. x 450.)

Recently chronic hepatitis has been defined as "inflammation of the liver continuing without improvement for at least 6 months" (74). Approximately 10% of patients develop symptomatic chronic viral hepatitis (47,60), of which two types are distinguished morphologically, namely chronic persistent and chronic active hepatitis. The classification of chronic hepatitis has been histological (5,63) because of the large variability in the clinical presentation and biochemical profiles, making these criteria unreliable for distinguishing between the forms of chronic hepatitis. Chronic persistent hepatitis is a benign disorder with a generally good prognosis, while chronic active hepatitis, the more severe disease, often progresses to cirrhosis and hepatic failure.

CHRONIC PERSISTENT HEPATITIS (CPH)

Chronic persistent hepatitis is the most common long-term sequel of acute hepatitis. It often presents with non-specific symptoms, principally fatigue. The levels of serum transaminases are slightly elevated and serum bilirubin is usually normal. The liver exhibits portal expansion by chronic inflammation, predominantly lymphocytic. Short septa may extend from them, but the architecture remains intact. The margins of the infiltrates and the limiting plate are sharp (Figure 9). There may be a variable degree of focal lobular inflammation and single cell necrosis. Evidence of chronic persistent hepatitis may remain as cases progress to chronic active hepatitis (11), and transition to cirrhosis is uncommon (4,68).

CHRONIC ACTIVE HEPATITIS (CAH)

Chronic active hepatitis occurs 2-3 times less often than chronic persistent hepatitis (60). This is not a single disease, but rather a pattern of progression of liver disease due to a variety of causes. HBV is the causative agent in from 10% to over 50% of cases. Geographic location is in part responsible for differences in the incidence of hepatitis B antigenemia in CAH (57,28). This incidence has ranged from 25% to 29% in studies from the United States (57,76). Acute hepatitis is seldom documented in these patients. Chronic liver disease is often first suspected because of the gradual onset of fatigue and anorexia, followed by the development of jaundice. It may be discovered incidentally, when liver function tests are performed for other medical reasons. Serum transaminases, AST (SGOT) and ALT (SGPT) are commonly elevated, serum globulins tend to be increased and

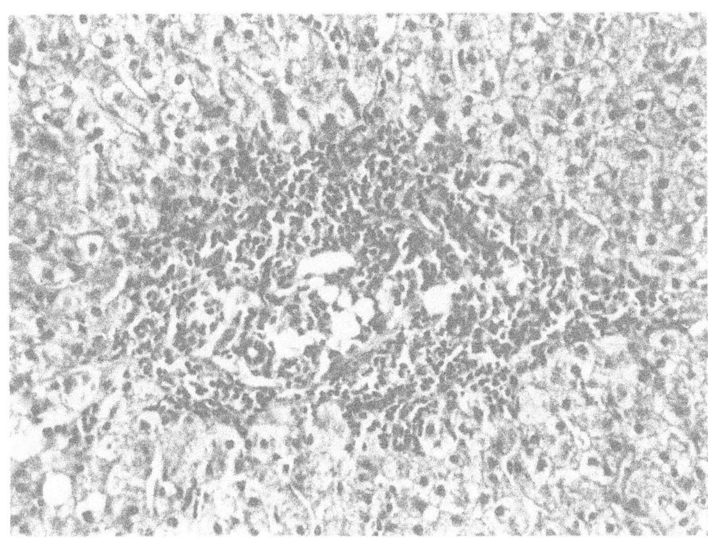

Figure 9. Chronic persistent hepatitis. Portal expansion by lymphocytic infiltrate. Short septa extend from portal tracts, but the limiting plate remains intact. (Hematoxylin and eosin stain. x 206.)

the bilirubin level is often high. Because of a generally unfavorable prognosis and the frequent institution of therapy in CAH, it is important not to overdiagnose this disease within a few months of the onset of acute hepatitis (23). The principal diagnostic histological features of CAH are portal inflammation and necrosis involving the periportal parenchyma (5,53). This is the process of continuing, band-like liver cell destruction at the interface of parenchyma and connective tissue, associated with a predominantly lymphocytic and plasma cell infiltrate. The slow progression of the inflammatory infiltrate from the portal tracts into the parenchyma results in an irregular and poorly visualized limiting plate (Figure 10). There is progressive deposition of new collagen in these areas, with formation of irregular fibrous septa at the periphery. The degenerating hepatocytes, often individually or in small groups, are entrapped and encompassed by the fibrous tissue of expanding portal areas. Destruction of periportal liver cell plates results in liver cell regeneration. Regenerating hepatocytes exhibit gland-like or "rosette"-like structures composed of more than three cells organized about a lumen (Figure 11). The cells forming the rosettes are either swollen and pale-staining or granular and eosinophilic, owing their appearance to an abundance of mitochondria (40).

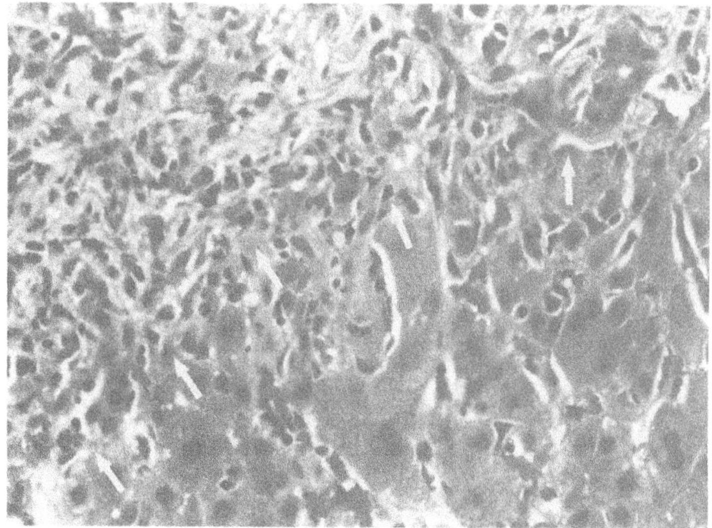

Figure 10. Chronic active hepatitis. The border between parenchyma and connective tissue (the limiting plate) is irregular (arrows) because of liver cell necrosis. Band-like liver cell destruction is associated with lymphocytic and plasma cell infiltrates. (Hematoxylin and eosin stain. x 450.)

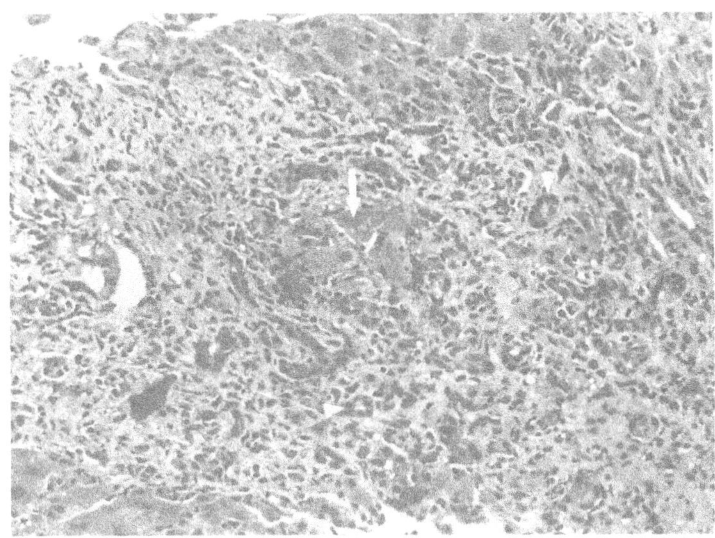

Figure 11. Severe, chronic active hepatitis. An enlarged portal area, with irregular fibrous septa at the periphery is shown. There is extensive loss of periportal parenchyma. Hepatocytes are entrapped within the fibrous tissue of the expanded portal tract (arrow). Regenerating hepatocytes form "rosette"-like structures (arrowheads). (Hematoxylin and eosin stain. x 206.)

In exacerbations of CAH acute intralobular changes are usually superimposed on chronic portal and periportal changes. Ballooning and acidophilic degeneration, focal necrosis and sometimes minimal to moderate cholestasis, with a proliferation of cholangioles, may be seen (55,3). Histologic cholestasis is not a usual feature of chronic hepatitis, and its presence may indicate bile secretory failure of liver cells in patients with hepatic failure.

Chronic active hepatitis is not a uniform process, as is acute viral hepatitis. The liver not only lacks the centrilobular predisposition to acute cells, but there is considerable variability from one lobule to the next in the distribution and extent of the inflammatory activity, necrosis and regeneration (48). With continuing development of fibrous septa and multiple regenerating parenchymal nodules, the effacement of normal liver architecture and transition toward cirrhosis begins. The inflammatory activity usually continues at the interface between fibrous septa and nodules (Figure 12). In the presence of clinical and functional exacerbation the diagnosis of "active cirrhosis" or "CAH with

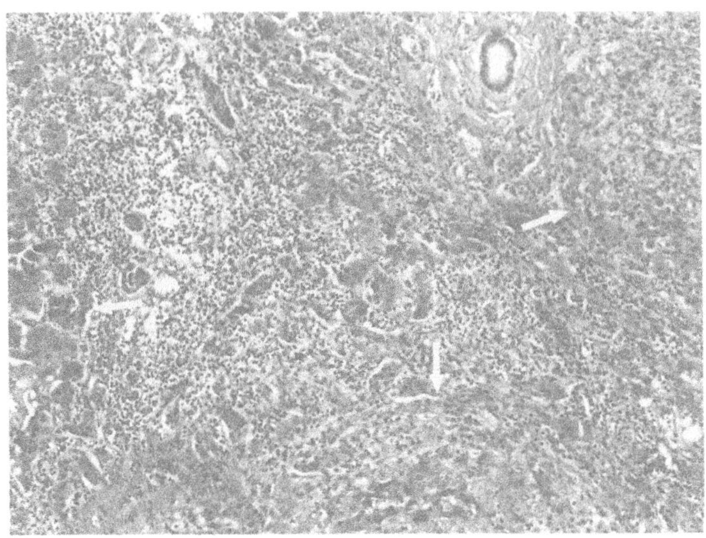

Figure 12. Active cirrhosis. Bridging necrosis involving portal and central area with formation of nodules (arrows). The intensive inflammatory activity is present at the interface between fibrous septa and nodules. (Hematoxylin and eosin stain. x 140).

transition to cirrhosis" is justified, even in the absence of discrete nodules entirely surrounded by fibrous tissue, that are typical for fully established cirrhosis. There is no difference between these two concepts, except in emphasis and indication of pathogenesis of the lesion.

Bridging hepatic necrosis may appear in more severe cases of chronic active hepatitis, when foci of periportal necrosis become confluent and form bridges (8). Portal-to-portal bridging at first and then, prognostically more important, portal-to-central bridging develop, and rapidly progress to cirrhosis (5) (Figure 12).

Damage of the epithelium of medium-sized bile ducts in CAH sometimes is demonstrated (55). Ductular epithelium exhibits mononuclear and neutrophilic infiltration, focal vacuolation, necrosis, and sometimes stratification (55,48,12). The importance of this feature is the association of a lesion with quicker progression to cirrhosis (55) and the necessity for its differentiation from primary biliary cirrhosis (13).

CAH is not a single disease but rather a complex continuous spectrum of a disease process of varying severity, that has different prognostic implications. A diagnosis of CAH is, therefore, insufficient by itself and needs to be qualified according to the severity.

Minimal or mild CAH is characterized by mild focal periportal necrosis and an intact hepatic architecture. It prognostically falls in the same category with chronic persistent hepatitis. The clinical course is often benign or very slowly progressive. CAH of moderate severity shows enough periportal necrosis to distort the lobular architecture. There is irregular portal fibrosis with some extension into the lobules. Clinically silent, slow progression to cirrhosis is common, although the lesion is potentially reversible (18). Severe CAH falls into the prognostic category of active cirrhosis. Bridging necrosis, reticulum collapse, and fibrosis play an important role in severe CAH. The changes are irreversible.

CIRRHOSIS

The frequency of cirrhosis after viral hepatitis has been reported to vary from less than 1% to 3% (6). Cirrhosis was recently defined by a WHO Committee as "a diffuse process characterized by fibrosis and the conversion of normal liver architecture into structurally abnormal nodules" (1). It is a chronic, progressive, irreversible condition that results in hepatic failure and portal hypertension. The most frequent type of cirrhosis following viral hepatitis is macronodular (3,19,34), although a mixed micro- and macronodular cirrhosis can develop. Cirrhosis following viral hepatitis is also called "postnecrotic," implying that it follows a single episode of acute hepatitis with massive necrosis. It is known now that survivors of fulminant hepatitis rarely develop cirrhosis, while most cases of cirrhosis develop from progression of chronic active hepatitis.

Macronodular cirrhosis is characterized by regenerating nodules of varying shape and size, usually 1-5 cm across (Figure 13). The fibrous septa are also irregular and often broad, following the collapse of several lobules (Figure 14). The newly formed regenerated nodules and vascular distortion produced by central-portal bridging leads to the development of portasystemic shunts and complete reconstruction of the original architecture of the liver. The end result of this continuous process is a small, scarred, coarsely nodular liver.

Dysplasia of liver cells may appear in the livers with macronodular cirrhosis, especially in hepatitis B virus carriers (2). This is expressed by cell enlargement, nuclear pleomorphism and hyperchromatism, and multinucleation. Dysplasia of hepatocytes may precede or coexist with liver cell carcinoma. However,

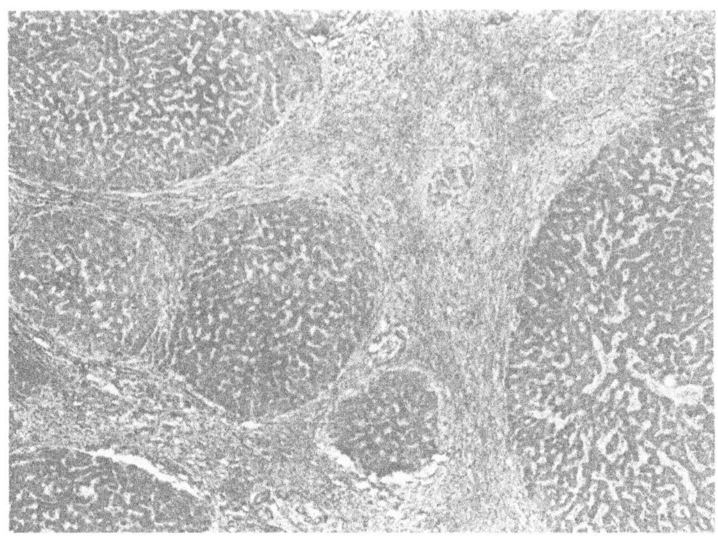

Figure 13. Cirrhosis. Regenerating nodules of varying shape and size are circumscribed by bands of connective tissue. (Hematoxylin and eosin stain. x 50.)

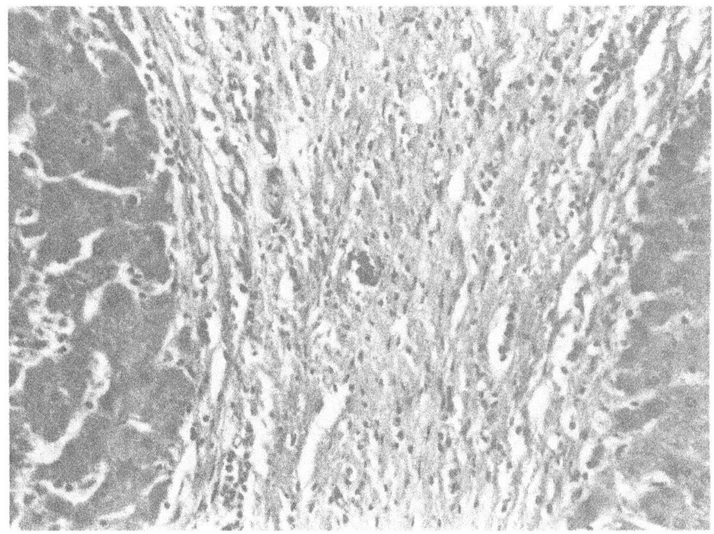

Figure 14. Cirrhosis. Broad fibrous septum, with mild inflammatory infiltrate, separates nodules. (Hematoxylin and eosin stain. x 206.)

the presence of dysplasia does not prove the existence of carcinoma in the liver and may persist for many years.

RELATION OF HB TO PRIMARY HEPATOCELLULAR CARCINOMA (PHC)

Relatively uncommon in the Western world (0.2-0.7%), PHC is common in Southeast Asia, Japan, Oceania, Greece and Italy (5.5%) (41) and is the most common form of cancer in certain parts of Africa (20%) (64,65). Five to twenty percent of the population of these countries are carriers of HBV. On the other hand, more than 90% of patients with PHC have HBsAg or high titers of anti-HBc associated with continued viral replication in their sera (44,45, 70). Even in areas with low incidence of hepatocellular carcinoma, serologic evidence of persistent infection with HBV is more common in patients with PHC than in the general population (7).

In most cases, PHC arises in a liver that exhibits evidence of chronic injury (Figure 15). Although it is more commonly associated with cirrhosis than with chronic hepatitis alone (22, 56), it also occurs in chronic HBV carriers in the absence of cirrhosis (66), suggesting a direct etiologic relationship between HBV and PHC.

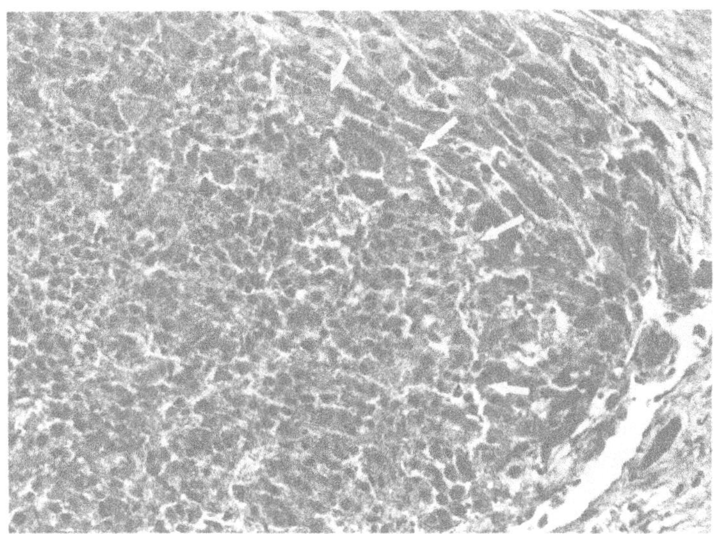

Figure 15. Primary hepatocellular carcinoma. PHC arising in cirrhotic nodule. The arrows point to the neoplasm. (Hematoxylin and eosin stain. x 206.)

The statistical association between chronic HBV infection and PHC, based on serologic studies, is supported by the common occurrence of HBV markers in liver tissue obtained at autopsies from patients with PHC. HBsAg and HBcAg are often found in the non-malignant hepatocytes adjacent to the tumor; these markers are more common in well-differentiated than in poorly differentiated hepatocytes (45,37,46,72,73,39). Both findings suggest that viral replication is better supported by normal hepatocytes or by neoplastic cells with minimal deviation from normal.

Based on the gross appearance, PHC is usually divided into solitary massive, nodular and diffuse infiltrative types (50). It tends to occur more often in the right lobe of the liver.

Solitary massive tumor may replace the entire lobe, infiltrating irregularly into the surrounding liver. It may outgrow its blood supply, resulting in hemorrhage and necrosis, particularly prominent in this form of PHC.

In the most frequent nodular type of PHC, multiple neoplastic nodules are scattered throughout the liver. They may coalesce and reach 5 cm in diameter but are usually discrete, gray-yellow, soft and (sometimes) cystic lesions. It is not known whether the neoplasm arises unicentrically and metastasizes intrahepatically or is multifocal in origin. The infrequent diffuse, infiltrative type of PHC spreads within fibrosis of a cirrhotic liver. Miriads of minute neoplastic nodules surrounded by collagen are sometimes indistinguishable from cirrhosis by gross inspection. The outstanding histological feature of PHC is the resemblance of the neoplastic cells to normal hepatocytes and the arrangement to trabeculae of normal liver. Depending on the degree of anaplasia, the tumor cells may be indistinguishable from normal and recognizable as a tumor only by growth pattern or may have an appearance of primitive cells with little resemblance to the hepatocytes.

The neoplastic hepatocytes retain the cohesiveness and ample eosinophilic cytoplasm, both of which are lost with increasing degree of anaplasia. The cytoplasm becomes polychromatophilic and basophilic, and less abundant. The nuclei tend to be larger, with prominent condensation of heterochromatin along the nuclear membrane. The nucleoli are larger than those in normal hepatocytes (Figure 16). Fat, bile and round hyaline bodies may be found in well-differentiated PHC. Alpha fetoprotein (AFP) appears in the sera and in the cytoplasm of moderately differentiated tumors. It is usually not associated with the extremes of well-differentiated and anaplastic PHC (50).

The less differentiated variants of PHC include anaplastic giant-cell forms, with multinucleated tumor cells in frequent mitosis, spindle-cell tumors and clear cell carcinomas (10).

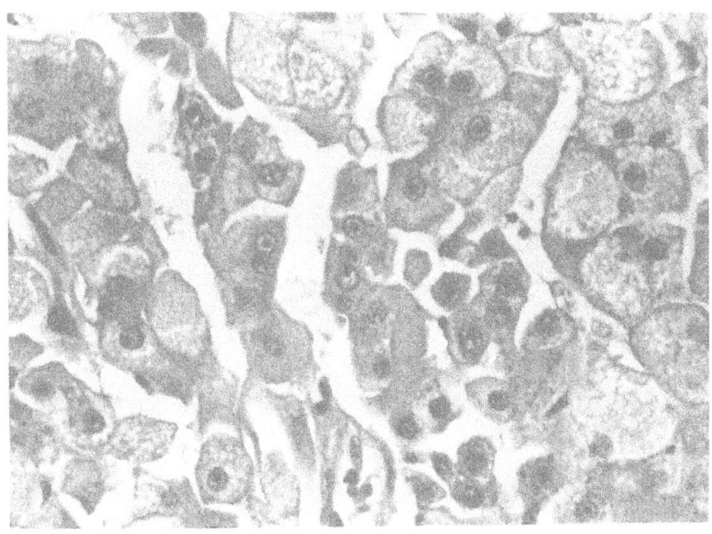

Figure 16. Primary hepatocellular carcinoma. Microtrabecular growth pattern simulates normal liver. A sinusoidal network is seen between the trabeculae. The neoplastic cells resemble normal hepatocytes. Small droplets of fat are seen in some cells. (Hematoxylin and eosin stain. x 450.)

There are several variations in the arrangement of the tumor cells in PHC (50). The microtrabecular growth pattern is the most common and occasionally simulates normal liver. The cells are arranged in cores 2-3 or more cells thick. A sinusoidal network rather than a connective tissue stroma is found between the trabeculae (Figure 16). The pattern is characterized as macrotrabecular when the cords of neoplastic cells reach 25-30 cells in thickness. Often cohesive sheets of tumor cells fit like a jigsaw puzzle and grow without pattern, giving a cobblestone appearance. Sometimes the neoplastic cells are arranged about a retained canaliculus, thus giving an acinar appearance. When tumor cells resume an elongated configuration with basally placed nuclei, the pattern of growth becomes pseudoglandular (Figures 17A, B). Rarely, a marked accumulation of secretion produces an adenoid pattern, sometimes with cystic areas. Areas of all of the above described patterns of growth may be seen in one tumor. The degree of anaplasia, as well as the pattern of growth does not influence the prognosis.

Primary hepatocellular carcinoma may be multifactorial in etiology. Aflatoxins and nitrosamines have been incriminated in the development of liver cancer. Some experimental evidence for

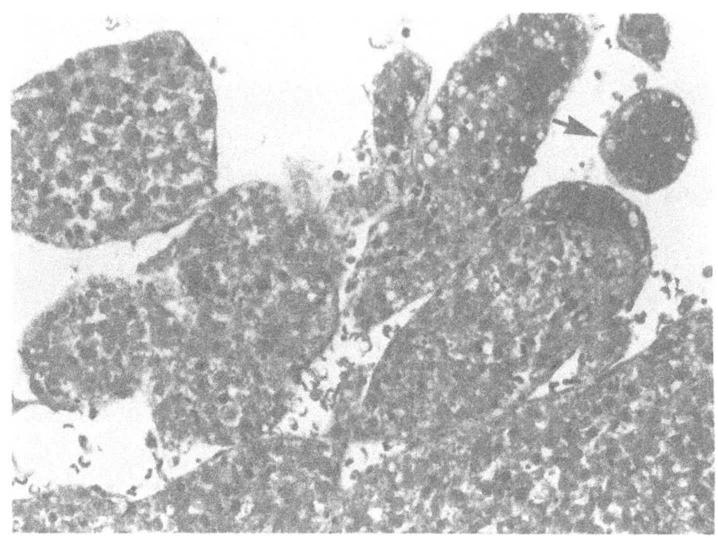

Figure 17A. Primary hepatocellular carcinoma, mixed pattern. Most of the tumor is represented by cohesive sheets of necrotic neoplastic cells. One group of neoplastic cells is arranged in pseudoacinar structure (arrow). (Hematoxylin and eosin stain. x 206.)

a synergism between HBV and various chemicals exists (27,42). The assumption that the virus and carcinogens act as co-carcinogens seems possible. Although conclusive evidence for the oncogenicity of the virus requires further proof, it is probable that chronic HBV infection is one of the important contributors to the development of PHC.

CONCLUSIONS

The morphologic appearance of hepatitis B and its sequelae are reviewed.

Depending on the intensity and virulence of the infection and the quality and quantity of the immune response, several clinical and pathological forms of HB exist. In those who develop acute viral hepatitis centrilobular hepatocytolysis, mononuclear inflammatory reaction and hepatocellular regeneration are the most characteristic features.

In more severe forms of acute viral hepatitis confluent necrosis involves central and midzones, with subsequent collapse of the reticulum framework and bridging formation. These

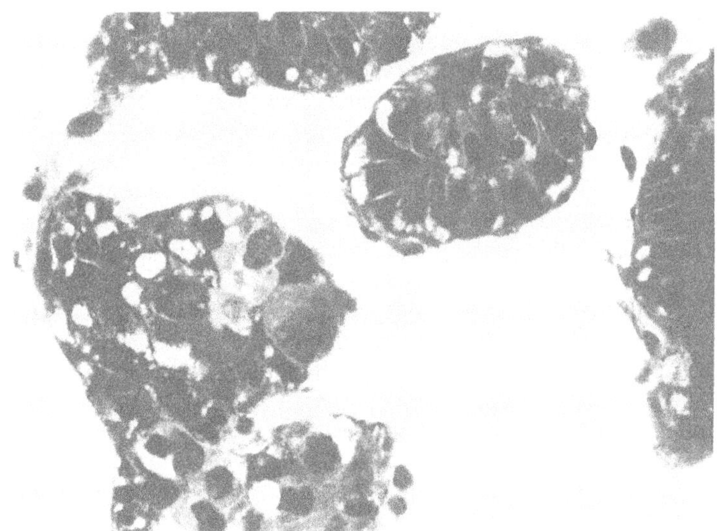

Figure 17B. Primary hepatocellular carcinoma. High power of A, showing pseudoglandular growth pattern of primary hepatocellular carcinoma. The neoplastic cells are elongated with basally placed nuclei. (Hematoxylin and eosin stain. x 450.)

patients tend to develop chronic liver disease. Less than 1% of patients develop the fulminant course of the disease, which is characterized by massive necrosis of all zones of the lobules and is associated with about a 90% mortality. Complete morphologic and clinical recovery is possible in both forms of severe acute viral hepatitis.

Hepatitis B may lead to a chronic HBsAg carrier state, with or without progression to chronic diseases; namely, chronic persistent hepatitis, chronic active hepatitis and cirrhosis.

Chronic HBV infection is associated with an increased incidence of primary hepatocellular carcinoma.

REFERENCES

1. Anthony, P. P., et al. (1977). The morphology of cirrhosis: definition, nomenclature and classification. Bull. WHO 55:521-540.
2. Anthony, P. P., Vogel, C. L., and Barker, L. F. (1973.) Liver cell dysplasia: a premalignant condition. J. Clin. Pathol. 26:217-223.

3. Baggenstoss, A. H., Summerskill, W. H. J., and Ammon, A. V. (1974). The morphology of chronic hepatitis, in The Liver and its Diseases, F. Schaffner, S. Sherlock and C. M. Leevy, eds., Intercontinental Medical Book Corporation, New York, pp. 199-206.
4. Becker, M. D., Scheuer, P. J., Baptista, A., and Sherlock, S. (1970). Prognosis of chronic persistent hepatitis. Lancet 1:53-70.
5. Bianchi, L., DeGroote, J., Desmet, V. J., Gedigh, P., Korb, G., Popper, H., Poulsen, H., Scheuer, P. J., Schmid, M., Tahler, H., and Welper, W. (1977). Acute and chronic hepatitis revisited. Lancet 2:914-919.
6. Bjorneboe, M. (1974). Viral hepatitis and cirrhosis. Clin. Gastroenterol. 3:409-418.
7. Blumberg, B. S. and London, W. T. (1980). Hepatitis B virus and primary hepatocellular carcinoma: Relationship of "Icrons" to cancer, in Viruses in Naturally Occurring Cancers, M. Essex, G. Todaro and H. zur Hausen, eds., Cold Spring Harbor Laboratory, New York, pp. 401-421.
8. Boyer, J. L. (1976). Chronic hepatitis. A perspective on classification and determinants of prognosis. Gastroenterology 70:1161-1171.
9. Boyer, J. L. and Klatskin, G. (1970). Pattern of necrosis in acute viral hepatitis. Prognostic value of bridging (subacute hepatic necrosis). N. Engl. J. Med. 283:1063-1071.
10. Buchanan, T. F., Jr. and Huvos, A. G. (1974). Clear cell carcinoma of the liver. Am. J. Clin. Pathol. 61:529-539.
11. Celle, G., Testa, R., Valpi, O., and Dodero, M. (1980). Morphologic evaluation of CAH with and without therapy. A two year follow-up. Hepato-Gastroent. 27:283-285.
12. Christofferson, P. and Dietrichson, O. (1974). Histological change in liver biopsies from patients with chronic hepatitis. Acta. Path. Microbiol. Scand. (A) 82:539-546.
13. Christofferson, P., Poulsen, H., and Scheuer, P. J. (1972). Abnormal bile duct epithelium in chronic aggressive hepatitis and primary biliary cirrhosis. Human Pathology 3:227-235.
14. Cook, G. C. and Sherlock, A. (1965). Jaundice and its relation to therapeutic agents. Lancet 1:175-195.
15. Chisari, F. V. and Edgington, T. S. (1975). Lymphocyte E rosette inhibitory factor: a regulatory serum lipoprotein. J. Exp. Med. 142:1092-1107.
16. Chisari, F. V., Routenberg, J. A., and Edgington, T. S. (1977). Mechanisms responsible for defective human T-lymphocyte sheep erythrocyte rosette function associated with hepatitis B virus infection. J. Clin. Invest. 57:1227-1238.

17. Chisari, F. V., Routenberg, J. A., Fiala, M., and Edgington, T. S. (1977). Extrinsic modulation of human T-lymphocyte E-rosette function associated with prolonged hepatocellular injury after viral hepatitis. J. Clin. Invest. 59:134-142.
18. DeGroote, J., Fevery, J., and Lepoutre, L. (1978). Long term follow-up of chronic active hepatitis of moderate severity. Gut. 19:510-513.
19. Desmet, V. J. (1972). Chronic hepatitis (including primary biliary cirrhosis), in The Liver, E. A. Gall and F. K. Mostofi, eds., Williams and Wilkins, Baltimore, pp. 286-341.
20. Desmet, V. J. and DeGroote, J. (1974). Histological diagnosis of viral hepatitis. Clin. Gastroenterol. 3: 337-354.
21. Edgington, T. S. and Chisari, F. V. (1975). Immunological aspects of hepatitis B virus infection. Am. J. Med. Sci. 270:213-227.
22. Edmondson, H. A. and Steiner, P. E. (1954). Primary carcinoma of the liver: a study of 100 cases among U.S. 900 necropsies. Cancer 7:462-503.
23. Fauerholdt, L., Asnaes, S., Ranek, L., Schigdt, T., and Tygstrup, N. (1977). Significance of suspected "chronic aggressive hepatitis" in acute hepatitis. Gastroenterology 73:639-645.
24. Galambos, J. T. (1964). Chronic persisting hepatitis. Am. J. Dig. Dis. 9:817-831.
25. Gerber, M. A. and Thung, S. N. (1979). The localization of hepatitis viruses in tissues. Int. Rev. Exp. Pathol. 2:49-76.
26. Gocke, D. J. (1978). Immune complex phenomena associated with hepatitis, in Viral Hepatitis, G. N. Vyas, S. N. Cohen and R. Schmid, eds., Franklin Institute Press, Philadelphia, pp. 227-283.
27. Gyorkey, F., Melnick, J. L., Mirkovic, R., et al. (1977). Experimental carcinoma of the liver in macague monkeys exposed to diethylnitrosamine and hepatitis B virus. J. Natl. Cancer Inst. 59:1451-1467.
28. Hadizyannis, S. J. (1974). Chronic viral hepatitis. Clin. Gastroenterol. 3:391-408.
29. Hadziyannis, S. J., Vissoulis, C., Moussauros, A., and Afroudakis, A. (1972). Cytoplasmic localization of Australia antigen in the liver. Lancet 1:976-979.
30. Hopf, U., Meyer zum Buschenfeld, K. H., and Arnold, W. (1976). Detection of a liver-membrane autoantibody in HBsAg-negative chronic active hepatitis. N. Engl. J. Med. 294:578-582.
31. Horney, J. T. and Galambos, J. T. (1977). The liver during and after fulminant hepatitis. Gastroenterology 73: 639-645.

32. Huang, S. N. (1975). Immunohistochemical demonstration of hepatitis B core and surface antigens in paraffin sections. Lab. Invest. 33:88-95.
33. Ishak, K. G. (1976). Light microscopic morphology of viral hepatitis. Am. J. Clin. Pathol. 65:787-827, 1976.
34. Ishak, K. G. (1972). Viral hepatitis. The morphologic spectrum, in The Liver, E. A. Gall and F. K. Mostofi, eds., Williams and Wilkins, Baltimore, pp. 218-268.
35. Kaff, R. S. (1978). Viral Hepatitis. Wiley, New York.
36. Karvountzis, G. G., Redeker, A. G., and Peters, R. L. (1974). Long-term follow-up of patients surviving fulminant viral hepatitis. Gastroenterology 67:870-877.
37. Kew, M. C., Ray, M. B., Desmet, V. J., et al. (1980). Hepatitis B surface antigen in tumor tissue and non-tumorous liver in black patients with hepatocellular carcinoma. Br. J. Cancer 40:399-406.
38. Krugman, S. and Giles, J. P. (1979). Viral hepatitis, type B (MS-2 Strain). Further observations on natural history and prevention. N. Engl. J. Med. 300:101-106.
39. Kubo, Y., Okuda, K., Musha, H., et al. (1978). Detection of hepatocellular carcinoma during a clinical follow-up chronic liver disease. Gastroenterology 74:578-582.
40. Lefkowitch, J. H., Arborgh, B. A. M., and Scheuer, P. G. (1980). Oxyphilic granular hepatocytes, mitochondrion-rich liver cells in hepatic disease. Am. J. Clin. Pathol. 74:432-441.
41. Lin, J. J. (1970). Primary cancer of the liver. Scand. J. Gastroenterol. 5 (Suppl. 6):223-241.
42. Lin, J. J., Liu, C., and Svoboda, D. J. (1974). Long term effect of aflatoxin B and viral hepatitis on marmoset liver. A preliminary report. Lab. Invest. 30:267-278.
43. MacSween, N. M. R. (1980). Pathology of viral hepatitis and its sequelae. Clin. Gastroenterol. 9:23-45.
44. Maupas, P., Larouze, B., London, W. T., Werner, B., Millman, I., O'Connell, A., Blumberg, B. S., Saimot, G., and Payet, M. (1975). Antibody to hepatitis B core antigen in patients with primary hepatic carcinoma. Lancet 2:9-11.
45. Nayak, N. C., Dhar, A., Sachdeva, R., Mittal, A., Seth, H. N., Sudarsanam, O., Reddy, B., Wagholikar, U. L., and Reddy, C. R. R. M. (1977). Association of human hepatocellular carcinoma and cirrhosis with hepatitis B virus surface and core antigens in the liver. Int. J. Cancer 20:643-654.
46. Nazarewicz, T., Krawezynski, K., Sluzarczyk, J., et al. (1977). Cellular localization of hepatitis B virus antigens in patients with hepatocellular carcinoma co-existing with liver cirrhosis. J. Infect. Dis. 135: 298-302.

47. Nielson, J. O., Dietrichson, O., Elling, P., and Christofferson, P. (1971). Incidence and meaning of persistence of Australia antigen in patients with acute viral hepatitis: development of chronic hepatitis. N. Engl. J. Med. 285:1157-1160.
48. Peters, R. L. (1975). Viral hepatitis: A pathologic spectrum. Am. J. Med. Sci. 270:17-31.
49. Peters, R. L., Omata, M., Ashcavai, M., and Liew, C. T. (1978). Protracted viral hepatitis with impaired regeneration, in Viral Hepatitis, G. N. Vyas, S. N. Cohen and R. Schmid, eds., Franklin Institute Press, Philadelphia, pp. 79-84.
50. Peters, R. L. (1976). Pathology of hepatocellular carcinoma, in Hepatocellular Carcinoma, K. Okuda, R. L. Peters, eds., Wiley, New York-London-Sydney-Toronto, pp. 107-168.
51. Phillips, M. J. and Paucell, S. (1981). Modern aspects of the morphology of viral hepatitis. Human Pathology 12: 1060-1084.
52. Popper, H. (1972). The pathology of viral hepatitis. Can. Med. Assoc. J. 106:447-452.
53. Popper, H., Paronetto, I., and Schaffner, F. (1965). Immune processes in the pathogenesis of liver disease. Ann. N. Y. Acad. Sci. 124:781-799.
54. Popper, H. and Udenfriend, S. (1970). Hepatic fibrossis. Am. J. Med. 49:707-721.
55. Poulsen, H. and Christofferson, P. (1972). Abnormal bile duct epithelium in chronic aggressive hepatitis and cirrhosis: a review of morphology and clinical, biochemical and immunologic features. Human Pathology 3: 217-225.
56. Prates, M. D. and Torres, F. D. (1965). A cancer survey in Lourenco Margues, Portugese East Africa. J. Natl. Cancer Inst. 35:729-757.
57. Prince, A. M. (1971). Role of serum hepatitis virus in chronic liver disease. Gastroenterology 60:913-921.
58. Rappaport, A. M. (1969). Anatomic considerations, in Diseases of the Liver, L. Schiff, P.I. J. B. Lippincott, Philadelphia.
59. Redeker, A. G. (1974). Fulminant hepatitis, in The Liver and its Diseases, F. Schaffner, S. Sherlock and C. M. Leevy, eds., Intercontinental Medical Book Corporation, New York, pp. 149-155.
60. Redeker, A. G. (1975). Viral hepatitis: clinical aspects. Am. J. Med. Sci. 270:9-16.
61. Robinson, W. S. (1978). Hepatitis B Dane particles DNA structure and the mechanisms of the endogenous DNA polymerase reaction, in Viral Hepatitis, G. N. Vyas, S. N. Cohen and R. Schmid, eds., Franklin Institute Press, Philadelphia, pp. 139-145.

62. Schaffner, F. (1966). Intralobular changes in hepatocytes and the electron microscopic mesenchymae response in acute viral hepatitis. Medicine 45:547-552.
63. Scheuer, P. J. (1977). Chronic hepatitis: A problem for the pathologist. Histopathology 1:5-19.
64. Schweitzer, T. L., Mosley, J. W., Ashcavai, M., et al. (1973). Factors influencing neonatal infection by hepatitis B virus. Gastroenterology 65:277-283.
65. Schweitzer, T. L., Wing, A., McPeak, C., and Spears, R. L. (1972). Hepatitis and hepatitis-associated antigen in 56 mother-infant pairs. JAMA 220:1092-1095.
66. Shikata, T., Yamazaki, S., and Uzawa, T. (1977). Hepatocellular carcinoma and chronic persistent hepatitis. Acta Pathologica Japonica 27:297-304.
67. Shouval, D. Chakraborty, P. R. Ruiz-Opazo, N., Baum, S., Spigland, T., Muchmore, E., Gerber, M. A., and Thung, S. N. (1980). Chronic hepatitis in chimpanzee carriers of hepatitis B virus: morphologic, immunologic and viral DNA studies. Proc. Natl. Acad. Sci. USA 77:6147-6151.
68. Sidi, S., Michel, H., Peters, R. L., and Redeker, A. G. (1974). Chronic persistent hepatitis. A clinicopathological study of 100 cases. Arch. Fr. Mal. App. Dig. 63:285-295.
69. Soloway, R. D., et al. (1972). Clinical, biochemical and histological remission of severe chronic active liver disease: a controlled study of treatments and early prognosis. Gastroenterology 63:820-833.
70. Tabor, E., Gerety, R. J., Vogel, C. L., Bayley, A. C., Anthony, P. P., Chan, C. H., and Barker, L. F. (1977). Hepatitis B virus infection and primary hepatocellular carcinoma. J. Natl. Cancer Inst. 58: 1197-1200.
71. Tisdale, W. A. (1966). Clinical and pathologic features of subacute hepatitis. Medicine 45:557-563.
72. Trevisian, A., Realdi, G., Losi, C., et al. (1978). Hepatitis B virus antigens in primary hepatic carcinoma: immunofluorescent techniques on fixed liver tissue. J. Clin. Pathol. 31:1133-1139.
73. Turbitt, M. L., Patrick, R. S., Gouldie, R. B., et al. (1977). Incidence in Southwest Scotland of hepatitis B surface antigen in the liver of patients with hepatocellular carcinoma. J. Clin. Pathol. 30:1124-1128.
74. U.S. Government Printing Office: Diseases of the Liver and Biliary Tract. Standardization of Nomenclature. Diagnostic Criteria, and Diagnostic Methodology. (1976). Washington, Fogarty International Center Proceedings No. 22 DHEW Publication No. (NIH) 76-725. DHEW, Washington, D. C.

75. Wands, J. R., et al. (1975). Cell-mediated immunity in acute and chronic hepatitis. J. Clin. Invest. 55: 921-929.
76. Wright, R., McCollum, R. W., and Klatskin, G. (1969). Australia antigen in acute and chronic liver disease. Lancet 2:117-121.

ASSAYS FOR HEPATITIS B VIRUS

Chung-Mei Ling

Abbott Laboratories
North Chicago, IL

Along the course of a typical hepatitis B virus (HBV) infection, three viral antigens appear in patients' serum: surface antigen (HBsAg), core antigen (HBcAg) and e antigen (HBeAg) (Figure 1). The host, in response to the infection, sooner or later produces corresponding antibodies, anti-HBs, anti-HBc and anti-HBe. All these antigens and antibodies have been used as diagnostic markers to determine the status of HBV infection: thus, HBsAg indicates current infection (2,17) while anti-HBs signifies immunity of the patient to HBV (5). The presence of detectable amounts of HBcAg (as Dane particles) (1,4) or HBeAg (12,13,14) reflects high concentrations of HBV and high infectivity; and anti-HBe, on the other hand, often coincides with lower levels of HBV and infectivity (Table 1). Anti-HBc (6,15) generally coexists with HBsAg or anti-HBs and therefore is a marker for current or past infection. When anti-HBc is the only HBV marker, it indicates one of two possibilities: 1) Current infection with HBV with HBsAg at a level too low to be detected; 2) Very early or very late convalescence with anti-HBs at a level too low for detection. A test for IgM class anti-HBc should be helpful to distinguish between early or late stage of convalescence (16). In HBsAg positives, the levels of the IgM antibody also distinguish recent infection or long-term chronic infection (Figure 2). Commercial diagnostic kits are available for all the above serological markers of HBV except for HBcAg.

Hepatitis B virus (the Dane particle) is known to consist of at least four components: HBsAg, HBcAg, a circular DNA and a DNA polymerase (7). Either the HBcAg or the DNA polymerase can be used as a marker for the presence of high levels of the virus. Epidemiological studies also exploit the subtypes of HBsAg, which can be determined by subtyping tests (11).

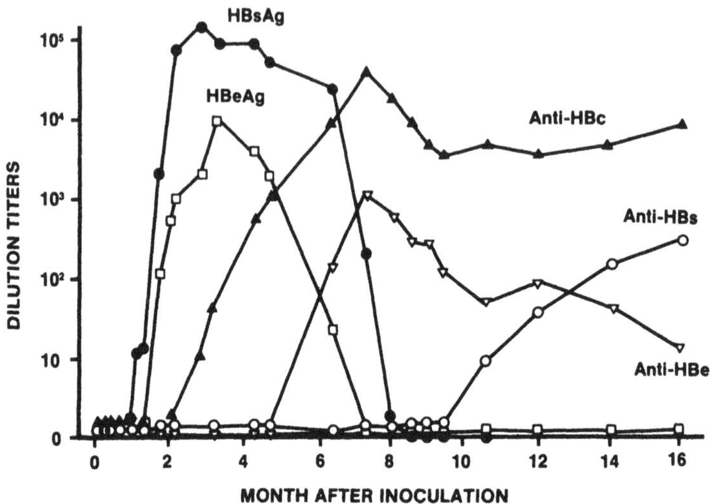

Figure 1. Time course of various markers of HBV infection in an experimental chimpanzee. The animal was inoculated with HBV on day 0.

ASSAY FOR HBsAg

For diagnostic purposes, HBsAg positive indicates current infection by HBV and therefore is unquestionably indicative of infectivity. Three types of test kits are now available commercially. They are radioimmunoassay (RIA), enzyme immunoassay (EIA) and reverse passive hemagglutination (RPHA). Both the RIA and EIA are solid-phase sandwich type as illustrated in Figures 3 and 4.

Table 1. Correlation of HBeAg, anti-HBe and HBcAg with HBsAg titer

HBsAg CEP titer	Total	No. of samples[a] Positive HBeAg	Positive anti-HBe	Positive HBcAg[b]
>1:64	6	6 (100)	0 (0)	6 (100)
1:32	5	3 (60)	2 (40)	2 (40)
1:16	24	3 (13)	20 (83)	1 (4)
1:8	17	1 (6)	16 (94)	0 (0)
<1:8	55	7 (13)	44 (80)	2 (4)
Total	107 (100)	20 (19)	82 (78)	11 (10)

[a]Percentages given in parentheses.
[b]All HBcAg positive samples were also HBeAg positive.

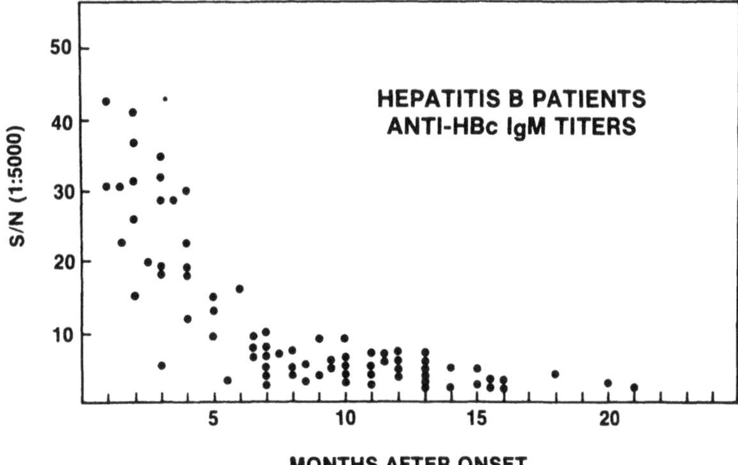

Figure 2. Levels of IgM class anti-HBc along the course of HBV infection.

They consist of an anti-HBs coated solid-phase which fishes out HBsAg in serum samples during an incubation period. After this, the solid-phase is exposed to a solution containing anti-HBs labelled with ^{125}I-iodine (RIA) or with an enzyme, horse-radish peroxidase (EIA). Again, some of the labelled anti-HBs molecules will be attracted to and attach to the solid-phase if it has been

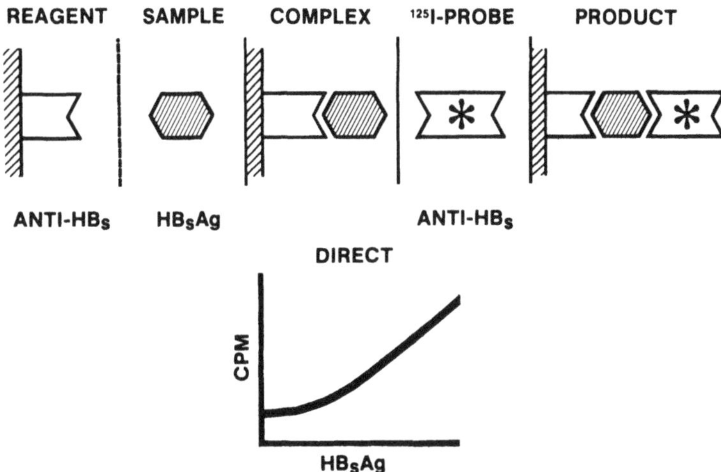

Figure 3. Schematic illustration of solid-phase sandwich radioimmunoassay for HBsAg. The curve at bottom indicates the relationship of uptake of radioactivity (cpm) by the solid-phase and the level of HBsAg present.

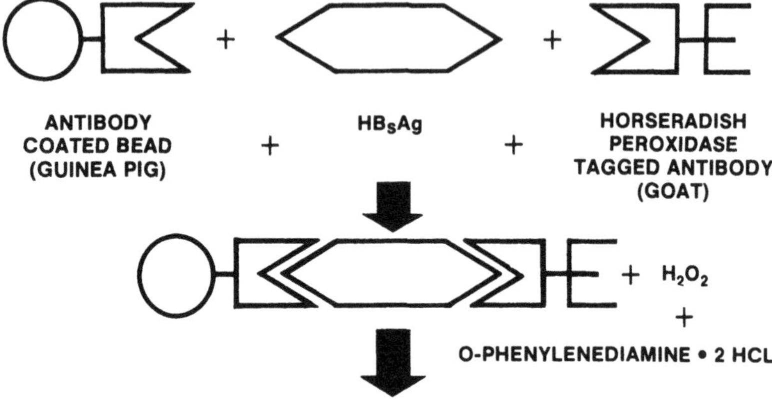

Figure 4. Schematic illustration of a solid-phase enzyme immunoassay for HBsAg (AUSZYME).

precoated with HBsAg in the first incubation. After thorough washing, the solid-phase is determined respectively for radioactivity or for enzyme activity with a suitable substrate such as o-phenylenediamine. Reverse passive hemagglutination test uses red blood cells coated with highly purified anti-HBs to test HBsAg in serum samples (Figure 5). The presence of HBsAg in serum will

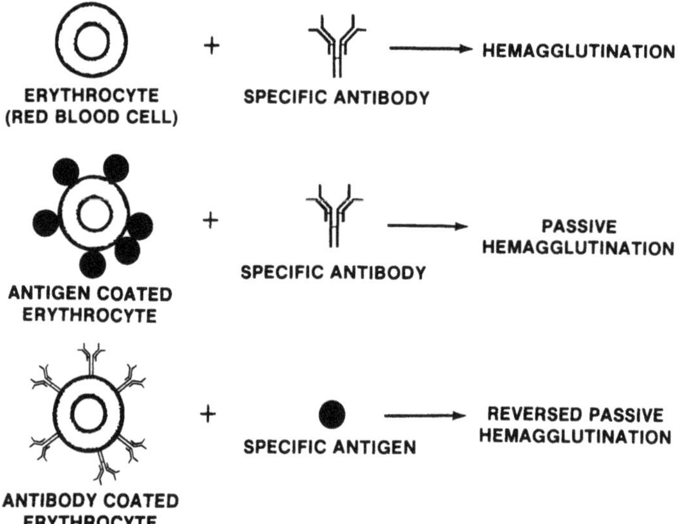

Figure 5. Schematic illustration of three types of hemagglutination. Reverse passive hemagglutination (RPHA) shown at the bottom is used for HBsAg detection.

cause the red blood cells to agglutinate. The latter, however, is not as sensitive as and less specific than the other two tests.

ASSAY FOR HBcAg AND HBeAg

The configurations of HBcAg and HBeAg test are essentially the same as that of the sandwich HBsAg tests except anti-HBc and anti-HBe respectively are used in the place of anti-HBs (13,15) (Figures 6 and 7). Both RIA and EIA kits are available commercially for HBeAg test but not for HBcAg test. For diagnostic purposes, both HBeAg and HBcAg positivity indicates high levels of HBV infection or high infectivity which often reflects poor prognosis (12,13,14). These two markers are generally quantitatively parallel to each other (Table 1), but HBcAg is much more difficult to detect than HBeAg. Test for HBeAg in serum does not require pretreatment of serum samples; while before performing HBcAg test (15), serum samples are first subjected to ultra-centrifugation to pellet the virus, which is then treated with a detergent and mercaptoethanol solution to uncoat the virus and to expose the core where HBcAg resides. Another reason that the ultra-centrifugation is necessary is the presence of high titer of anti-HBc in the serum of HBV infected patients (10,11). Anti-HBc interferes with HBcAg test and must be removed by means of ultra-centrifugation. HBcAg test has not been available in kit form. It remains to be a research tool only.

ASSAYS FOR ANTI-HBs

Anti-HBs is the only marker that indicates immunity of the patient to HBV, and, unless coexists with HBsAg, noninfective for the virus. Until very recently, solid-phase sandwich radioimmunoassay (5) has been dominating the diagnostic scene of anti-HBs. An EIA kit recently entered the field and terminated the RIA monopoly. These two tests, nearly identical in performance, are almost complimentary templates of the sandwich RIA and EIA for HBsAg. Solid-phase is coated with HBsAg, and ^{125}I-labelled HBsAg (RIA) or enzyme labelled HBsAg (EIA) is used in the liquid reagent. During the first incubation period (overnight), the solid-phase is used to fish out anti-HBs in a serum sample. After a washing step, the liquid reagent is added to allow the formation of the other side of the sandwich of antigen-antibody-antigen structure. The solid-phase is then measured for radioactivity or enzyme activity.

ASSAYS FOR ANTI-HBc AND ANTI-HBe

Both anti-HBc and anti-HBe are markers for HBV infections, current or past. They are most easily assayed with solid-phase RIA

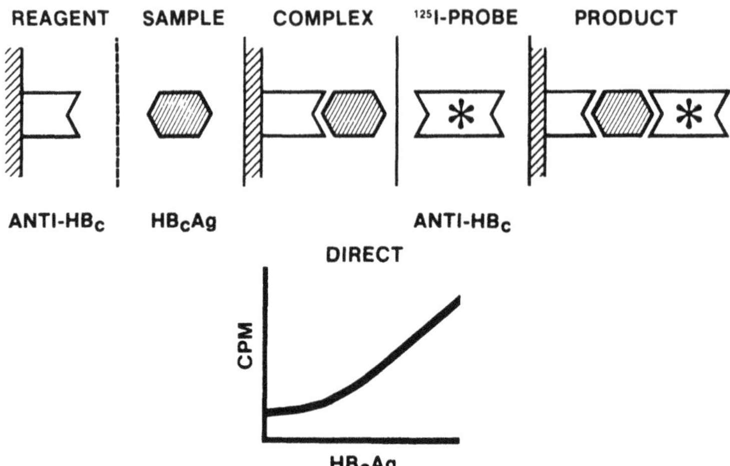

Figure 6. Schematic illustration of solid-phase sandwich radioimmunoassay for anti-HBc. See also Figure 3 legend.

or EIA of competitive mode (15) (Figure 8). The commercial kits for anti-HBc employ HBcAg coated 6 mm polystyrene beads as the solid-phase and ^{125}I- or enzyme-labelled anti-HBc IgG in a one-step incubation with a serum sample. After overnight incubation, the bead is washed and determined for ^{125}I-radioactivity or enzyme activity respectively. As illustrated in Figure 8, if anti-HBc is present in the sample, the labelled and the unlabelled anti-HBc compete for the limited quantity of HBcAg on the solid-phase,

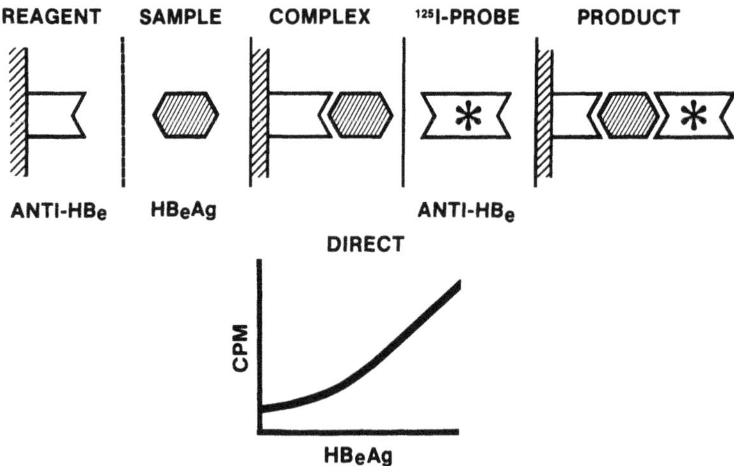

Figure 7. Schematic illustration of solid-phase sandwich radioimmunoassay for HBeAg. See also Figure 3 legend.

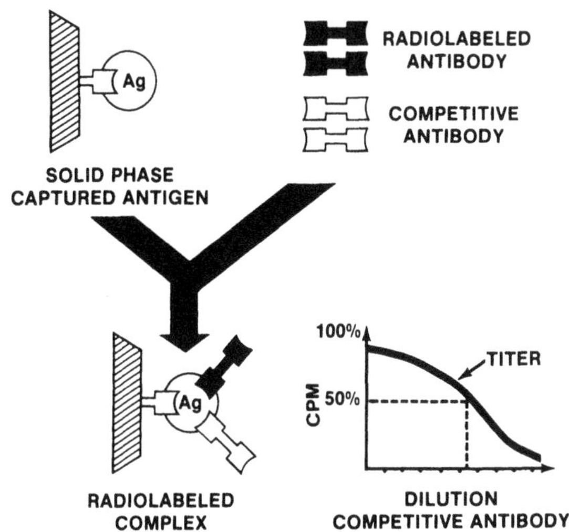

Figure 8. Competitive solid-phase radioimmunoassay. The curve at lower right shows the relationship of uptake of radioactivity by the solid-phase and antibody levels.

which results in less binding of labelled antibody. Thus, the higher the anti-HBc in the sample, the lower the signal measured. It is therefore a competitive assay.

The commercial test kit for anti-HBe is a neutralization test (13). An aliquot of sample is mixed with equal volume of HBeAg solution of precalibrated potency and, during an incubation period, the HBeAg will be neutralized by anti-HBe, if present, in the sample. The mixture is then tested for HBeAg by the sandwich method described before (Figure 7). Like the anti-HBc assay, when anti-HBe is present in the sample, neutralization of HBeAg results in lower signal. The neutralization test used for anti-HBe detection is a more sensitive test than the competition test used for anti-HBc detection. In general, the anti-HBe levels in patients are much lower than the anti-HBc levels. We therefore chose the more sensitive neutralization test for anti-HBe.

OTHER ASSAYS OF IMPORTANCE

Figure 9 shows an artist's rendering of hepatitis B virus, which consists of the coat or surface antigen (HBsAg), the core (HBcAg), the DNA and the enzyme DNA polymerase. It has been suggested (19) that HBeAg is a cryptic component of HBcAg. Besides the tests for the coat and the core described before, assaying the DNA polymerase, the structure of the DNA and the antigenic subtypes of HBsAg are also of diagnostic and scientific values. Since these

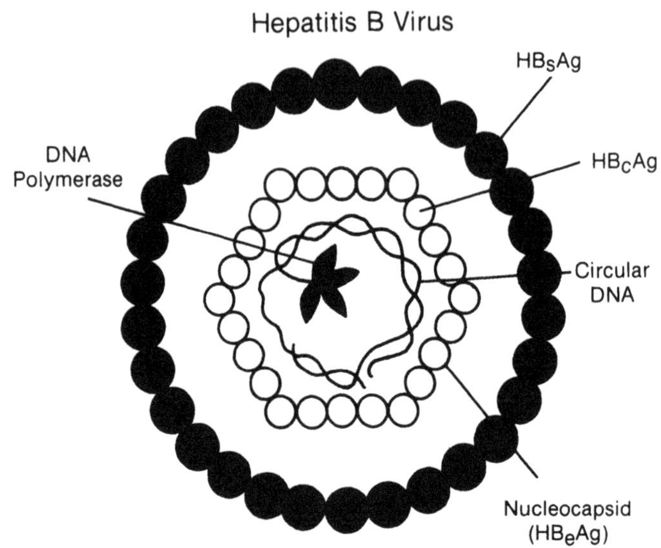

Figure 9. Schematic illustration of the structure of a hepatitis B virus.

tests are not available in convenient commercial kit form, they remain to be research tools in some research oriented laboratories. I will only briefly describe the methods to perform these tests.

Subtyping of HBsAg and anti-HBs: the easiest way to do subtype determinations is by immunodiffusion method. When anti-HBs of AD subtype is placed in the center well, an ad subtype antigen shows a spike over an ay subtype antigen in the peripheral wells. If purified anti-HBs of D subtype (anti-D) or of Y-subtype (anti-Y) is in the center well, only ad subtype or ay subtype antigen respectively will show precipitin lines (11). We radiolabelled purified anti-D or anti-Y and used them in the place of the labelled antibody of AUSRIA test; this then became a radioimmunoassay for HBsAg subtyping. Likewise, r and w subtypes can be determined by using labelled anti-R and anti-W subtype antibodies respectively. Anti-HBs subtypes can be determined by specific inhibition by special subtypes of HBsAg prior to anti-HBs test.

DNA POLYMERASE ASSAY

DNA polymerase associated with HBV can be determined by a standard DNA polymerase assay measuring the incorporation of radioactive nucleotide into acid insoluble fractions (6). Serum samples are usually concentrated 50 fold for this test. A word of caution for using this test to assess HBV: samples contaminated with bacteria also give positive DNA polymerase test.

HBV DNA DETERMINATION

HBV DNA in a sample can be determined by DNA hybridization technique using radiolabelled HBV DNA fragments as the probe (8). Recently, recombinant cloned HBV DNA has been used for the detection of HBV in serum by DNA hybridization technique (3,18). The method was said to be more sensitive than the DNA polymerase and HBeAg assays in detecting the presence of HBV particles. At least one company sells reagents needed for HBV DNA determination (BRL).

REFERENCES

1. Almeida, J. D., Rubenstein, D., and Scott, E. J. (1971). New antigen-antibody system in Australia antigen positive patients. Lancet 2:1225.
2. Blumberg, B. S., Gerstley, B. J. S., Hungerford, D. A., London, W. T., and Sutnick, A. I. (1967). A serum antigen (Australia antigen) in Down's syndrome, leukemia and hepatitis. Ann. Intern. Med. 66:924.
3. Chakraborty, P. R., Ruiz-Opazo, N., Shouval, D., and Shafritz, D. A. (1980). Identification of integrated hepatitis B virus DNA and expression of viral RNA in a HBsAg producing human hepatocellular carcinoma cell line. Nature 286:531.
4. Dane, D. S., Cameron, C. H., and Briggs, M. (1970). Virus-like particles in serum of patients with Australia antigen-associated hepatitis. Lancet 1:695.
5. Ginsberg, A. L., Conrad, M. E., Bancroft, W. H., Ling, C. M., and Overby, L. R. (1973). Antibody to Australia antigen: detection with a simple radioimmunoassay, incidence in military populations and role in the prevention of hepatitis B with gammaglobulin. J. Lab. Clin. Med. 82:317.
6. Greenman, R. L., Robinson, W. S., and Vyas, G. N. (1975). A sensitive test for antibody against the hepatitis B core antigen. Vox Sang. 29:77.
7. Hoofnagle, J. H., Gerety, R. J., Ni, L. Y., and Barker, L. F. (1974). Antibody to hepatitis B core antigen. A sensitive indicator of hepatitis B virus replication. N. Engl. J. Med. 290:1336.
8. Hung, P. P., Ling, C. M., Mao, J. C.-H., and Overby, L. R. (1975). Hepatitis B antigen: molecular hybridization between free DNA from plasma and Dnae particle DNA. Nature 253:571.
9. Ling, C.-M. (1980). Radioimmunoassays for hepatitis B virus markers, in Manual of Clinical Immunology, 2nd edition, N. R. Rose and H. Friedman, eds., Am. Soc. Microbiol., Washington, D. C., p. 372.

10. Ling, C.-M., Mushahwar, I. K., Overby, L. R., Berquist, K. R., and Maynard, J. E. (1979). Hepatitis B e-antigen and its correlation with other serological markers in chimpanzees. Infect. Immun. 24:352.

11. Ling, C.-M., Irace, H., Decker, R. H., and Overby, L. R. (1973). Hepatitis B virus antigen: validation and immunologic characterization of low-titer serums with ^{125}I-antibody. Science 180:203.

12. Magnius, L. O. and Espmark, J. A. (1972). New specificities in Australia antigen positive sera distinct from LeBouvier determinants. J. Immunol. 109:1017.

13. Mushahwar, I. K., Overby, L. R., Frosner, G., Deinhardt, F., and Ling, C.-M. (1978). Prevalence of hepatitis B e-antigen and its antibody as detected by radioimmunoassays. J. Med. Virol. 2:77.

14. Nielson, J. O., Deitrichson, O., and Juhl, E. (1974). Incidence and meaning of the "e" determinant among hepatitis B antigen positive patients with acute and chronic liver diseases. Lancet 2:913.

15. Overby, L. R. and Ling, C.-M. (1976). Radioimmune assay for anti-core as evidence for exposure to hepatitis B virus. St. Lukes Med. Bull. 15:83.

16. Overby, L. R., Ling, C.-M., Decker, R. H., Mushahwar, I. K., and Chau, K. (1982). Serodiagnostic profiles of viral hepatitis, in Viral Hepatitis, 1981 International Symposium, W. Szmuness, H. J. Alter and J. E. Maynard, eds., Franklin Institute Press, Philadelphia, pp. 169.

17. Prince, A. M. (1968). An antigen detected in the blood during the incubation period of serum hepatitis. Proc. Natl. Acad. Sci. USA 60:814.

18. Weller, I. V. D., Fowler, M. J. F., Monjadino, J., and Thomas, H. C. (1982). The detection of HBV-DNA in serum by molecular hybridization: a more sensitive method for the detection of complete HBV particles. J. Med. Virol. 9:273.

19. Yoshizawa, H., Ito, Y., Simonetti, J. P., Takahashi, T., Miyakawa, Y., and Mayumi, M. (1979). Demonstration of hepatitis B e antigen in hepatitis B core particles obtained from the nuclei of hepatocytes infected with hepatitis B virus. J. Gen. Virol. 42:513.

IMMUNOLOGIC RESPONSES TO HEPATITIS B VIRUS AND THEIR INTERPRETATIONS

Barbara G. Werner, Jules L. Dienstag, Barbara J. Kuter, David R. Snydman, B. Frank Polk, Donald E. Craven, Richard Platt, Clyde S. Crumpacker and George F. Grady

From the Boston Inter-Hospital Hepatitis B Vaccine Study Group (State Laboratory Institute, Department of Public Health, Commonwealth of Massachusetts; Massachusetts General Hospital; Tufts-New England Medical Center; Brigham and Women's Hospital; Channing Laboratory; Boston University Hospital; Boston City Hospital; New England Deaconess Hospital; and Beth Israel Hospital) and the Departments of Medicine, Tufts University School of Medicine, Harvard Medical School and Boston University School of Medicine, Boston, Massachusetts; and Merck, Sharp and Dohme Research Laboratories, West Point, Pennsylvania

INTRODUCTION

The variable patterns of clinical and serologic responses following exposure to hepatitis B virus (HBV) have been well documented (1,2,3,4). The common serologic profiles are charted using the markers listed in Table 1.

The virus particles associated with HBsAg reactivity appear in three forms as demonstrated by electron microscopy. The complete virus is a 42 nm diameter double-shelled structure bearing HBsAg on the surface; it has an internal core with an immunologically distinct specificity, HBcAg. The nucleocapsid core contains a small partially double-stranded DNA and an associated DNA polymerase. Excess surface glycoprotein makes up the bulk of the circulating antigen which is in the form of small spherical particles, approximately 20 nm in diameter, and tubular or rod-like particles of the same diameter and variable in length. HBsAg as detected on these particles is seen during both acute and chronic infection with HBV.

Table 1. Terminology

HBV	Hepatitis B virus, formerly the Dane particle
HBsAg	Hepatitis B surface antigen, formerly Australia antigen (Au) or hepatitis associated antigen (HAA)
HBcAg	Hepatitis B core antigen
HBeAg	Hepatitis B e antigen
Anti-HBs	Antibody to HBsAg
Anti-HBc	Antibody to HBcAg
Anti-HBe	Antibody to HBeAg

Because the core antigen, HBcAg, is covered by HBsAg when the virus is in the circulation, free HBcAg is not routinely detected in the serum.

The combinations of these antigens and antibodies most frequently observed are shown in Table 2. The interpretations listed come largely from studies of sequential samples from individuals exposed to HBV. The other hepatitis B associated antigen, HBeAg, is found only in the presence of HBsAg; in both acute hepatitis B and in chronic HBsAg carriers, it is a good indicator of increased infectivity. Antibody to HBeAg, anti-HBe, is found along with HBsAg in resolving cases of acute hepatitis B and also in some HBsAg chronic carriers. Anti-HBe is also seen in combination with anti-HBc and anti-HBs and occasionally along with only one of these antibodies.

Sequential observations reveal three basic patterns of response. The first is associated with typical acute hepatitis B. The initial marker seen two to six weeks following exposure is HBsAg, followed shortly thereafter by HBeAg and then by anti-HBc. HBeAg decreases in concentration and generally becomes undetectable before HBsAg is lost (one to six months after appearance). Anti-HBs appears during convalescence; but sometimes it is not detected immediately after the loss of HBsAg, resulting in the so-called anti-HBc "window period." Anti-HBs, anti-HBc and anti-HBe may persist for many years. In a second pattern, HBsAg persists for longer than six months along with anti-HBc and with either HBeAg or anti-HBe; anti-HBs is not detected. This represents the HBsAg chronic carrier state. In another group of individuals HBsAg may never be detected, but anti-HBc, anti-HBs and anti-HBe can be observed and these antibodies persist.

The advent of a new hepatitis B vaccine, Heptavax-B (Merck, Sharp and Dohme, West Point, PA), has brought renewed interest in the interpretation and significance of the presence of these

Table 2. Most common combinations of serologic markers of HBV

Serologic Reactivity			
HBsAg	Anti-HBs	Anti-HBc	Interpretation
+	-	-	Likely to be early hepatitis B (HB)
+	-	+	Acute HB or chronic HB (carrier state)
-	+	+	Recovery from HB
-	+	-	Immunization with HBsAg or long after recovery from HBV infection
-	-	+	The "window" period, i.e. after loss of HBsAg and before detectable anti-HBs, or possibly long after recovery from HBV infection

serologic markers both for evaluating the need for vaccine and for studying subsequent responses to the vaccine. The vaccine is prepared from the excess surface coat of the virus, the small 20 nm particles which are devoid of HBV DNA. Administration of the vaccine induces antibody to HBsAg only. Thus, anti-HBs in the absence of anti-HBc and anti-HBe is observed in vaccinated individuals.

MATERIALS AND METHODS

As part of a trial of the efficacy and immunogenicity of the vaccine in health care personnel in six Boston hospitals, 2109 sera were screened for HBsAg, anti-HBc and anti-HBs to identify staff members eligible for participation. Of these, 1330 seronegative individuals subsequently received either vaccine or placebo (5).

The presence of the serologic markers was ascertained using commercial radioimmunoassays (RIA), AUSRIA-II for HBsAg, CORAB for anti-HBc, and AUSAB for anti-HBs (Abbott Laboratories, N. Chicago, IL). Calculation of the estimated RIA units as suggested by the manufacturer was performed to aid in quantitative evaluation of the levels of anti-HBs present.

RESULTS AND DISCUSSION

Screening and follow-up of these high-risk health care personnel revealed different immunologic responses, including anamnestic responses and borderline levels of antibodies which were difficult to evaluate.

After excluding individuals with a prior history of hepatitis B or with recent passive immunization with globulin, almost 14% of those screened for hepatitis B susceptibility were found to have HBV serologic markers. The marker seen most frequently was anti-HBs, which was detected in 12.6% of the samples tested. It was the only evidence of prior HBV exposure in 6.0% of the total; anti-HBc alone (0.7%) or HBsAg (0.5%) were found infrequently. Thus, the screening data indicate that anti-HBs should be the test of choice in determining past HBV exposure among health care workers.

The presence of both anti-HBs and anti-HBc was correlated with high levels of these antibodies. In general, lower levels of antibody were seen if only anti-HBc or anti-HBs were observed. The lowest levels of anti-HBs have S/N ratios of 2.1 - 9.9 (sample counts per minute/counts per minute of the negative control mean). Questions have been raised about the specificity and protective value of low levels of anti-HBs (6). While our data suggest that anti-HBs should be the test of choice in health care personnel, caution should be exercised in assessing immunity of those with low anti-HBs levels. For example, in the group followed after receiving placebo injections, 14 individuals had low levels of anti-HBs detected only intermittently. These study participants had detectable anti-HBs on from one to five different occasions which was not attributed to administration of hyperimmune globulin and which did not persist (Table 3). The same samples were again positive upon retesting but the levels remained low. One individual was positive on five of seven occasions but never had anti-HBs at a level substantially above the cutoff. The recommendations of the Immunization Practices Advisory Committee (7) to use a cutoff of S/N $\geq$ 10 to exclude individuals from vaccination programs should be adhered to. In addition, further studies on the specificity and protection afforded by the presence of low levels of antibody should be carried out.

During the course of the study, we also observed a number of individuals with low level anti-HBc reactivity. The assay for anti-HBc is a competitive RIA with $\geq$ 50% inhibition, indicating positivity. Those with low levels or borderline anti-HBc results made up a group exhibiting one of three basic patterns observed over time and shown in Table 4. In the first category, containing the majority of individuals observed in either vaccine or placebo recipients, the samples were consistently negative with percent inhibitions generally < 30%. Study participants 1 and 2 reflect

Table 3. Nonimmunized health care workers who acquire anti-HBs during 12-18 months of observation by serial bleeding

	No. individuals		
Months post randomization	Total tested	Anti-HBs(+)	Anti-HBs(+) not attributable to HBIG or clinical HB[a]
Prebleed	664	0	0
0	93	0	0
1	659	7	6
3	656	6	4
6	650	16	9
9	647	11	5
12	533	12	5
15	332	8	4
18	61	0	0

[a]Represents 14 different individuals with from one to five positive samples.

this pattern. The second group, represented by 3 and 4, contains actual cases of HBV infection who clearly seroconvert to an anti-HBc positive status, generally with a rapid rise to 90-100% inhibition in the assay. The third group includes study participants 5 and 6 who were in this class for over a year of follow-up (A) and individuals 7 and 8 who moved into the borderline category and did not continue to higher levels of anti-HBc (B). The individuals with clearly negative anti-HBc results and those with levels bordering on positivity lack other HBV markers; whereas those in category 2 exhibit one of the patterns of serologic markers discussed earlier. Whether these antibodies are specific for HBcAg or are cross reacting antibodies, this reactivity may appear in persons previously lacking it and the low levels may persist. Further studies are needed to further characterize these phenomena.

Only a small group, about 3%, did not respond to three 20 μg doses of the vaccine (5). At the other extreme were 36 individuals representing 5.6% of the vaccinees who had a high level of anti-HBs after only a single dose of vaccine (Table 5). As a group, they continued to have the highest titers of anti-HBs for the duration of the trial. Their response was consistent with an anamnestic response rather than the primary response expected from seronegative individuals. Long-term follow-up will show whether or not their immunity will persist longer than that of those exhibiting a more typical primary response. Additional studies are in progress to assess the response to a single dose of vaccine in individuals with low levels of anti-HBs (8).

Table 4.

Months post randomization	Representative patterns of anti-HBc results % inhibition[a]							
	Negative		"Classic" seroconversion		Borderline			
					A		B	
	1	2	3	4	5	6	7	8
Prebleed	0	26	2	10	45	38	0	10
0	-	-	1	-	45	-	-	4
1	9	18	67	10	42	40	9	0
3	9	8	86	21	46	42	41	9
6	24	5	91	24	50	33	38	0
9	4	0	94	1	31	47	58	5
10	-	-	-	-	-	-	44	-
12	21	33	98	98	44	32	53	59
13	-	-	-	-	-	-	50	46
15	2	22	95	100	54	58	-	49
18	18	0	-	-	29	20	-	-

[a]% inhibition = $$\frac{\text{Mean neg. control cpm} - \text{Mean sample cpm}}{\text{Mean neg. control cpm} - \text{Mean pos. control cpm}} \times 100$$

for serial samples from 8 different study participants representative of the three groups.

Table 5. Anti-HBs level one month after first dose of vaccine

Est. RIA units	No.	% of vaccinees
< 8[a]	278	41.7
8 - 512	351	52.7
> 512[b]	37	5.6
TOTALS	666	100.0

[a]Approximately equivalent to an S/N of 2.1 - 4.0.

[b]Approximately equivalent to an S/N > 100.

CONCLUSION

In addition to classic responses to infection and immunization, the screening and follow-up of vaccine and placebo recipients in our trial of hepatitis B vaccine also revealed different immunologic responses and borderline levels of antibodies. The findings demonstrate a new level of complexity in the interpretation of serologic markers associated with HBV. In addition, our data support the adoption of an anti-HBs assay, with a cutoff distinctly above marginal values, as a screening test to identify susceptible health care personnel prior to vaccination.

ACKNOWLEDGMENTS

We gratefully acknowledge the assistance of the laboratory staff, Kathryn Cyr and Robert Salman; the study coordinator, Rita Ouellet-Hellstrom; the study nurses, Susan O'Rourke, Eloise Watkins, Kathy Holbrook, Susan Fischer, Joanna Stull, Laurie Kunches, Patricia Berdine, Susan Marino and Jocelyn Loftus; the data processing and statistical center staff, Alvaro Munoz, Barbara Nash and Maureen Mahoney.

REFERENCES

1. Hoofnagle, J. H. (1981). Seminars in Liver Disease 1:7.
2. Seeff, L. B. (1981). Internal Medicine 2:41.
3. Mushahwar, I. K., Dienstag, J. L., Polesky, H. F., et al. (1981). Am. J. Clin. Pathol. 76:773.
4. Overby, L. R., Ling, C.-M., Decker, R. H., Mushahwar, I. K., and Chau, K., in Viral Hepatitis, 1981 International Symposium, W. Szmuness, H. J. Alter and J. E. Maynard, eds., Franklin Institute Press, Philadelphia, pp. 169-182.
5. Dienstag, J. L., Werner, B. G., Polk, B. F., et al. (1982). Hepatology 2:696. Abstract.
6. Nath, N., Fang, C. T. and Dodd, R. Y. (1982). Transfusion 22: 300.
7. Centers for Disease Control. (1982). MMWR 31:317.
8. Werner, B. G., Dienstag, J. L., Kuter, B. J., et al. (1983). Hepatology 3:878. Abstract.

HEPATITIS B INFECTION CONTROL AMONG PHYSICIANS, DENTISTS AND LABORATORY PERSONNEL

William J. Schneider

Albert Einstein College of Medicine
Bronx, New York
and
Montefiore Hospital
Bronx, New York

In 1949, Liebowitz and his coworkers provided the first description of a health worker, in this instance a blood bank technician, who developed hepatitis B as a consequence of her employment (1). Their observations on hepatitis B as a hazard to health care personnel have been extended by many others, especially after the emergence of new medical techniques, such as hemodialysis, which introduced into the hospital setting a new source of infection in the form of frequently transfused chronic renal disease patients who manifested a high prevalence of hepatitis B surface antigen (HBsAg) carriage (2,3,4). Other developments, including increasing parenteral drug abuse, the migration to the United States of population groups from areas of high hepatitis B endemicity, and disease transmission among male homosexuals also may have contributed to greater probability of exposure of health care workers to the hepatitis B virus. As a consequence, it is now well recognized that the effective control of this problem is of major importance. In defining a suggested approach to this problem, I will describe the extent to which physicians, dentists and laboratory personnel are subject to the risk of hepatitis B as well as the specific risk factor involved, the real and possible outcomes of infection, and, finally, the potentialities and shortcomings of the environmental, immunological, epidemiological and clinical modalities of infection control.

HEPATITIS B IN PHYSICIANS, DENTISTS AND LABORATORY PERSONNEL

Serological evidence of prior hepatitis B infection can be determined with assays for HBsAg, antibodies to the surface antigen (anti-HBs) and antibodies to the core antigen (anti-HBc), collectively referred to as markers of hepatitis B. Anti-HBs determinations alone will detect the vast majority of those who have experienced an infection in the past. Multiple studies comparing the prevalence of these markers among volunteer blood donors with that among physicians have revealed that the frequency of previous infection is two to four times greater in the physician groups (15 to 20% seropositivity) than among the population as a whole (5,6,7,8). Again, in comparison with randomly selected blood donors, evidence of prior hepatitis B in physicians increases with age or years of practice, suggesting a continuing risk of exposure to the virus (7). However, the rate of exposure appears to be greatest during the early years of clinical activity which would include hospital-based postgraduate training (7). In our own hospital experience, 10 of the 11 cases of hepatitis B in physicians during a consecutive four-year period occurred in physicians-in-training (4).

Similarly, dentists have been found to have had a history of clinical hepatitis almost three times as frequently as a comparable group of attorneys (9) and serological evidence of prior hepatitis infection two to three times that of the general population (10,11,12). Again, the prevalence of serologic markers among dentists increases with age or years of practice (11,12) and is greater among oral surgeons than among general dentists (9,12). Furthermore, HBsAg carriage among dental professions exceeds that in the population as a whole (11,12).

Clinical laboratory personnel share the risks of hepatitis B infection and seropositivity. Seventeen percent of 70 laboratory technicians in a midwestern hospital were found to be marker positive (13), while 23% of 85 personnel in an urban hospital in the northeast had evidence of prior hepatitis B (8). Only 5% of a control population of blood donors was positive in the second study.

In Liebowitz' early description of hepatitis B in a hospital employee, he suggested that contact with blood was an important determining factor (1). Indeed, this has been confirmed in a recent study by Dienstag and Ryan, and exposure to blood is undoubtedly the single most significant factor in hepatitis B acquisition (8). In contrast, patient contact is relatively unimportant. Other risk factors reflect the likelihood of blood contact or of contact with the clinical specimens of patients who themselves are prone to the HBsAg carrier state. Thus, surgeons and pathologists are more likely to be infected than are their nonsurgical colleagues both in medicine and dentistry (7,8,9,12); house officers are more vulnerable than more senior physicians (4,7). Employment in dialysis units, emergency wards, blood banks, operating rooms, pathology suites and clinical laboratories, dental facilities, and intensive care units is associated with higher frequencies of hepatitis B infection than is that in other clinical areas (8,14,15,16,17).

OUTCOMES OF HEPATITIS B INFECTION

Hepatitis B infection of health personnel has both personal and societal consequences. Although most cases are asymptomatic, each infected individual may be subject to the prolonged morbidity of acute hepatitis to which a very small percentage will succumb. Approximately 10% of those who develop hepatitis B will proceed to chronic active hepatitis with the ultimate development of cirrhosis in many. An association between prolonged HBsAg carriage and hepatocellular carcinoma has been noted (18). Prolonged HBsAg carriage may also occur in otherwise asymptomatic individuals.

The knowledge that a health care professional is HBsAg positive may also have significant emotional impact. More importantly, although very infrequently, this may lead to the transmission of the illness by the practitioner to his patients. Nine reports have appeared in the recent medical literature describing such events, involving dental professionals, physicians, a nurse and an inhalation therapist (19,20,21,22,23,24,25,26,27). It should be noted that some of these index cases had acute hepatitis B or were in the incubation period of the illness and that the number of episodes of transmission is very small, relative to the number of HBsAg positive health workers. Authorities in the field do not recommend practice restrictions for HBsAg carriers who have not been responsible for transmission (28,29). Nonetheless, the very contemplation of such a possibility is an additional source of concern for the individual HBsAg carrier and for society.

The overall economic impact of hepatitis B in health personnel is difficult to calculate. However, Ehrenkranz noted in 1975 that this illness accounted for 5% of all disabilities, 11% of lost work days, 21% of disability costs and 80% of non-trauma-related workers' compensation claims by hospital workers in one state (30). In our hospital's experience with hepatitis B in health professionals, the average duration of symptoms was 49 days with a range of 21 days to 210 days, and the average separation from employment was 51 days with a range of 12 to 215 days (4). It can easily be surmised that interruption of professional practices plus the cost of medical attention can be substantial.

PREVENTION STRATEGIES

With the commercial introduction of hepatitis B immune globulin (HBIG) in 1977, attention focused on the post-exposure prevention of this illness. In an effort to predict the efficacy of this approach at our hospital, we reviewed our experience with employees who developed hepatitis B during the years immediately prior to and following HBIG availability. Although a vigorous publicity campaign emphasized the needs and benefits of reporting relevant accidents to our Employees' Health Service and this resulted in 1118 reports of accidental needle punctures over a four-year period, only 4 of 30 employees who subsequently developed work-related hepatitis B sought medical attention in advance. Only 5 others could identify potentially responsible incidents in retrospect (4). Similar experiences have been recorded by others, noting a 6 to 25% awareness of a specific precipitating event under these circumstances (3,31,32,33).

Thus, effective prevention strategies must rely on those steps which would either prevent exposure to the causative agent and/or immunize vulnerable subpopulations against the illness. Environmental control measures, as refined over the course of the last decade, have already contributed meaningfully to some reduction in the incidence of hepatitis B among laboratory workers in particular (34). Specific recommendations are detailed in several publications (35,36,37).

Several of the more important concepts can be summarized. Known hazards, such as clinical specimens from HBsAg positive patients, should be identified as such in order to protect casual contacts. Known HBsAg positive patients should be clearly identified as well. By working cleanly and carefully, one can minimize the splattering of infectious fluids and accidental injuries. Needle punctures remain a source of concern and can be minimized by avoiding the recapping of used needles and by the prompt and appropriate disposal of soiled needles and sharp instruments (38). Disposable gloves should be employed when handling potentially contaminated specimens or when working in a hyperendemic area such as a dialysis unit. It has been demonstrated that viral antigen can remain on countertops and equipment for prolonged periods (35); therefore, any spills of specimens should be cleaned immediately and thoroughly. Disinfection and sterilization practice should take into account published recommendations which delineate the efficiency of boiling or autoclaving, ethylene oxide, glutaraldehyde, sodium hypochlorite and iodophors and the relative inadequacy of alcohols, phenolics and quarternary ammonium compounds (35). Eating, drinking and smoking in clinical areas increases the possibility of self-contamination and should be avoided. Pipetting by mouth should be discouraged for the same reason. Clinical specimens should be disposed in a manner which will assure that the unaware will not be placed in jeopardy. Finally, all clinical specimens, whether identified as originating from patients with hepatitis B or not, should be regarded as sources of potential contagion and should be handled carefully. Testing for HBsAg is not routine, and most carriers are not recognized as such when hospitalized (39,40). Additionally, other illnesses such as non-A, non-B hepatitis may be transmitted by blood, and caution is advised for this reason as well. Careful hand-washing after contact with a patient or clinical specimens is of fundamental concern.

Additional enrivonmental control recommendations have been generated for specific clinical areas such as dialysis units (36).

Despite general acknowledgment of the environmental control measures cited above, hepatitis B remains a problem. In emergency situations, precautions tend to be modified and some personnel are less attentive to details of this nature than are others. Hepatitis B vaccine, which became commercially available in 1982, should prove to be the most effective control mechanism against this illness. In healthy adults, including health workers, the vaccine has been demonstrated to be safe and very effective with a 96-98% antibody response rate after the third 20 microgram dose (41,42). The vaccine is recommended for health care workers at risk (43); in the context of the observations cited above, medical and dental students, those practicing physicians and dentists whose responsibilities include repetitive exposure to blood, and clinical laboratory personnel who handle blood specimens would all fall into this category. Possible constraints to vaccination include the cost of the vaccine, concern about possible adverse consequences of vaccination and the anticipation of second generation vaccine refinements.

In order to fully evaluate the relative importance of vaccine expense, it must be compared with the total costs of the illness as a work-related hazard. Certain aspects of the financial and personal impact of hepatitis B on the individual, the health facility and society were addressed above. Furthermore, a recent cost analysis study suggested that the vaccination of high-risk groups such as surgical house officers is actually less expensive than the monitoring and the potential treatment costs of nonvaccination (44).

The safety of the hepatitis B vaccine can be surmised from the fact that over 20,000 individuals have been vaccinated with minimal immediate side effects, if any, and with no long-term reactions (45). The method of vaccine purification, including biophysical procedures and formalin, urea, and pepsin treatment inactivate all known viruses. Furthermore, safety testing of each vaccine lot in the laboratory and in chimpanzees should assure that the vaccine remains free of significant biologic hazard. While the prospects for a second generation synthetic hepatitis B vaccine appear bright (46,47), it will be a matter of years before such preparations are commercially available and the present danger to vulnerable personnel of contracting hepatitis B renders the use of the licensed vaccine advisable.

The question may be raised whether to test vaccine recipients for evidence of prior hepatitis experience before vaccine administration. In qualifying the response to this question, it should first be observed that no adverse health consequences were identified following the administration of vaccine to HBsAg positive or anti-HBs positive individuals (48). Thus, a decision can be reached solely on the basis of economics, taking into account the prevalence of hepatitis B serologic markers in a given subpopulation, the cost of testing, and the cost of the vaccine. Based upon these considerations, screening has been suggested to become cost-effective when a prevalence of seropositivity exceeds 10%, provided that the determination will cost no more than $15.00. For routine screening of this sort, only one test, anti-HBs or anti-HBc, need be employed. Thus, serologic screening would appear to be worthwhile or equivocal for practicing physicians, dental professionals, and laboratory workers, but not cost-effective for entry level personnel such as medical and dental students.

When a health worker who has not been vaccinated or who is not known to be immune to hepatitis B experiences a significant accidental exposure to hepatitis B such as a needle puncture or mucous membrane contact with known antigen-positive body fluids, passive immunization should be implemented. Hepatitis B immune globulin with an anti-HBs titer of 1:100,000 does appear to be more effective than placebo immune globulin (49). However, other studies comparing the more expensive hepatitis B immune globulin with currently available standard immune globulin which now does contain low titers of anti-HBs have yielded equivocal results (50). Nonetheless, it is currently recommended that significant hepatitis B exposures to susceptible individuals be managed with hepatitis B immune globulin at a dosage of 0.06 ml/kg immediately or within 24 hours of contamination, with a second like dosage to be administered one month later (51). Those individuals experiencing similar accidents, possibly but not definitely contaminated by HBsAg positive fluid, may be given standard immune globulin in the same pattern as above or they may not be treated at all. Accidental punctures or exposures from known HBsAg negative patients require no treatment.

It has been determined that equivalent antibody responses are generated by the hepatitis B vaccine, whether administered together with hepatitis B immune globulin or alone (52). In view of this, a good opportunity to administer the vaccine to personnel clearly in need of immunization, who have overlooked that need beforehand, may be in conjunction with the management of a provocative incident. This experience should render the individual all the more receptive to this intervention.

When all else fails, hepatitis B results. With this realization, until vaccination becomes standard and universal, it remains necessary to periodically screen health personnel in high-risk areas, dialysis units in particular, for HBsAg (36). This practice has relevance in terms of individual disease detection and is useful in monitoring environmental control measures. In our hospital 5 of 39 cases of hepatitis B manifested over a four-year period were identified by these means (4).

It is also important that those responsible for the medical attention of health care personnel--and physicians, dentists and laboratory workers themselves--be aware of the diverse clinical presentations of this illness. In our experience, health workers found to have hepatitis B were symptomatic for periods extending up to 20 days or more, averaging one week, prior to diagnosis. The most common presenting symptoms in these individuals were arthralgia, skin eruption or pruritis and fatigue or malaise. It is not surprising that the initial misdiagnosis of rheumatoid arthritis, photosensitivity and viral syndrome in 3 of 39 cases resulted from these nonspecific presentations (4).

SUMMARY

The attack rate of hepatitis B among physicians, dentists and laboratory personnel, as well as among several other categories of health workers, exceeds that of the general population. While contact with blood is the most significant common denominator in determining risk, the actual mode of transmission in individual episodes is usually unrecognized, and the disease itself is more often subclinical than not. However, the consequences of infection can be physically, to some extent emotionally, and financially devastating. In order to be effective, prevention strategies must focus on pre-exposure interventions because of the subtlety of transmission. Appropriate environmental safeguards are helpful and should be implemented and enforced. Hepatitis B vaccine offers the greatest promise of bringing this problem under control and should be made available to those health professionals at risk as early in their careers as possible.

REFERENCES

1. Liebowitz, S., Greenwald, S. L., Cohen, I., and Litwins, J. (1949). JAMA 140:1331.
2. Trumbull, M. L. and Geiner, D. J. (1951). JAMA 145:965.
3. Williams, S. V., Huff, J. C., Feinglass, E. J., Gregg, M. B., Hatch, M. H. and Matsen, J. M. (1974). Am. J. Med. 57:904.
4. Schneider, W. J. (1979). J. Occup. Med. 21:807.
5. Lewis, T. L., Alter, J. H., Chalmers, T. C., Holland, P. V., Purcell, R. H., Alling D. W., Young, D., Frenkel, L. D., Lee, S. L., and Lamson, M. E. (1973). N. Engl. J. Med. 289:647.
6. Smith, J. L., Maynard, J. E., Berquist, K. R., Doto, I. L., Webster, H. M., and Sheller, M. J. (1976). J. Infect. Dis. 133:705.
7. Denes, A. E., Smith, J. L., Maynard, J. E., Doto, I. L., Berquist, K. R., and Finkel, A. J. (1978). JAMA 239:210.
8. Dienstag, J. L. and Ryan, D. M. (1982). Am. J. Epidemiol. 115:26.
9. Feldman, R. E. and Schiff, E. R. (1975). JAMA 232:1228.
10. Glenwright, H. D., Edmondson, H. D., Whitehead, F. I. H., and Flewett, T. H. (1974). Brit. Dent. J. 136:409.
11. Mosley, J. W., Edwards, V. M., Casey, G., Redeker, A. G., and White, E. (1975). N. Engl. J. Med. 293:729.
12. Weil, R. B., Lyman, D. O., Jackson, R. J., and Bernstein, B. (1977). N. Y. State Dent. J. 43:587.
13. Levy, B. S., Harris, J. C., Smith, J. C., Washburn, J. W., Mature, J., Davis, A., Crosson, J. T., Polesky, H., and Hanson, M. (1977). Am. J. Epidemiol. 106:330.
14. Maynard, J. E. (1981). Am. J. Med. 70:439.
15. Pattison, C. P., Boyer, K. M., Maynard, J. E., and Kelly, P. C. (1974). JAMA 230:854.
16. Snydman, D. R., Bregman, D., and Bryan, J. A. (1977). J. Infect. Dis. 135:687.
17. Pantelick, E. L., Steere, A. C., Lewis, H. D., and Miller, D. J. (1981). Am. J. Med. 70:924.
18. Blumberg, B. S. and London, W. T. (1981). N. Engl. J. Med. 304:782.
19. Garibaldi, R. A., Rasmussen, C. M., Holmes, A. W., and Gregg, M. B. (1972). JAMA 219:1577.
20. Levin, M. L., Maddrey, W. C., Wands, J. R., and Mendeloff, A. I. (1974). JAMA 228:1139.
21. Snydman, D. R., Hindman, S. H., Wineland, M. D., Bryan, J. A., and Maynard, J. E. (1976). Ann. Intern. Med. 85:573.
22. Goodwin, D. J., Fannin, S. L., Roberto, R. R., Edwards, V. M., and Mosley, J. W. (1976). Gastroenterology 71:908. (Abstract.)

23. Rimland, D., Parken, W. E., Miller, G. B., Jr., and Schrack, W. D. (1977). N. Engl. J. Med. 296:953.
24. Collaborative Study by Central Public Health Laboratories. (1980). Lancet 1:1.
25. Grob, P. J., Bischof, B., and Naeff, F. (1981). Lancet 2: 1218.
26. Carl, M., Francis, D. P., Blakey, D. L., and Maynard, J. E. (1982). Lancet 1:731.
27. Reingold, A. L., Kane, M. A., Murphy, B. L., Checko, P., Francis, D. P., and Maynard, J. E. (1982). J. Infect. Dis. 145:262.
28. Alter, H. J., Chalmers, T. J., Freeman, B. M., Lunceford, J. L., Lewis, T. L., Holland, P. V., Pizzo, P. A., Plotz, P. H., and Meyer, W. J. (1975). N. Engl. J. Med. 292:454.
29. Alter, H. J. and Chalmers, T. J. (1981). Hepatology 1:467.
30. Ehrenkrantz, N. J. (1975). Arch. Environ. Health 30:514.
31. Bryan, J. A., Carr, H. E., and Gregg, M. B. (1973). JAMA 223:279.
32. Hansen, J. P., Falconer, J. A., Hamilton, J. D., and Herpok, F. J. (1981). J. Occup. Med. 23:338.
33. Callender, M. E., White, Y. S., and Williams, R. (1982). Br. Med. J. (Clin. Res.) 284:324.
34. Grist, N. R. (1981). Med. Lab. Sci. 38:103.
35. Bond, W. W., Petersen, N. J., and Favero, M. S. (1977). Health Lab. Sci. 14:235.
36. Center for Disease Control: Hepatitis--Control measures for hepatitis B in dialysis centers. (1977). Viral Hepatitis Investigations and Control Series.
37. National Health and Medical Research Council. (1981). Aust. Dent. J. 26:49.
38. McCormick, R. D. and Maki, D. G. (1981). Am. J. Med. 70: 928.
39. Linnemann, C. C., Jr., Hegg, M. E., Ramundo, N., and Schiff, G. M. (1977). Am. J. Clin. Pathol. 67:257.
40. Mahoney, J. P., Richman, A. V., and Teague, P. O. (1978). South. Med. J. 71:624.
41. Szmuness, W., Stevens, C. E., Harley, E. J., Zang, E. A., Oleszko, W. R., William, D. C., Sadovsky, R., Morrison, J. M., and Kellner, A. (1980). N. Engl. J. Med. 303:833.
42. Szmuness, W., Stevens, C. E., Harley, E. J., Zang, E. A., Taylor, P. E., Alter, H. J., and the Dialysis Vaccine Trial Group. (1981). J. Med. Virol. 8:123.
43. A.C.I.P. (1982). MMWR 31:317.
44. Mulley, A. G., Silverstein, M. D., and Dienstag, J. L. (1982). N. Engl. J. Med. 307:644.
45. Inter-Agency Group. (1982). MMWR 31:465.
46. Prince, A. M., Ikram, H., and Hopp, T. P. (1982). Proc. Natl. Acad. Sci. U.S.A. 79:579.

47. Young, P., Vandin, M., Dixon, J., and Zuckerman, A. J. (1982). J. Virol. Methods 4:177.
48. Dienstag, J. L., Stevens, C. E., Bhan, A. K., and Szmuness, W. (1982). Ann. Intern. Med. 96:575.
49. Seeff, L. B., Wright, E. C., Zimmerman, H. J., Alter, H. J., Dietz, A. A., Felsher, B. F., Finkelstein, J. D., Garcia-Pont, P., Gerin, J. L., Greenlee, H. B., Hamilton, J., Holland, P. V., Kaplan, P. M., Kiernan, T., Koff, R. S., Leevy, C. M., McAuliffe, V. J., Nath, N., Purcell, R. H., Schiff, E. R., Schwartz, C. C., Tamburro, C. H., Vlahcevic, Z., Zemel, R., and Zimmon, D. S. (1978). Ann. Intern. Med. 88:285.
50. Seeff, L. B. and Hoffman, J. H. (1979). Gastroenterology 77:161.
51. A.C.I.P. (1981). MMWR 30:423.
52. Szmuness, W., Stevens, C. E., Oleszko, W. R., and Goodman, A. (1981). Lancet 1:575.

CONTROL MEASURES FOR HEPATITIS B PROBLEMS IN DENTISTRY

Vernon J. Brightman and Robert Weibel

University of Pennsylvania School of Dental Medicine
Clinical Research Center
Philadelphia, Pennsylvania
and
The Joseph Stokes, Jr. Research Institute of the
Children's Hospital of Philadelphia
Philadelphia, Pennsylvania

INTRODUCTION

Infection with hepatitis B virus is a recognized occupational hazard of dentistry (1). There is considerable anecdotal evidence to support this; for example, most dentists, even if they have not been infected themselves, personally know other dentists or dental assistants who have been ill with this infection. The occupational nature of hepatitis B virus infection in dental personnel is also supported by the greater frequency of antibody to hepatitis B among dentists (and physicians) as compared with lawyers (professionals of similar socioeconomic status without occupational clinical contact), and by the gradual increase of the frequency of hepatitis B antibody in dentists (and physicians) to a maximum of about 20% over the normal span of dental and medical practice (2). Among dentists, the highest frequency of hepatitis B antibody occurs in oral surgeons and periodontists, with lower frequencies among general practitioners and pedodontists. The annual percentage increment of new infections for dentists and physicians is about 0.5, the increment for blood donors from the general population being about 0.087.

Reports of fatal fulminant infection with hepatitis B in dentists appear to be rare, and the major hazard for dental personnel apart from illness and time lost convalescing from an acute infection is the possibility of developing a chronic infection with this virus. Only a small minority of infections with hepatitis B result in chronic infection, and chronic infection with this virus in itself does not prevent a dentist from practicing. However, reports of a small number of instances in which dentists infected with hepatitis B have transmitted the infection to patients (3,4), and have lost their right to practice, have caused much concern among the dental profession. Concern about other complications of chronic infection, such as progressive liver damage and development of hepatomas, has been less apparent. This level of anxiety among dentists about occupationally acquired hepatitis B infection occurs in all western countries and has been accompanied by publication of professional bulletins and articles alerting dentists to the higher prevalence of hepatitis B infection among various groups of patients, e.g. S. E. Asian immigrants (5), hemodialysis patients and male homosexuals. The American Dental Association (6), the National Health Service in the United Kingdom (7) and other professional groups have issued recommendations for the dental management of patients with hepatitis B infection, and much of the effort of infection control committees in dental schools and hospitals in recent years has been concerned with protocols for screening patients for hepatitis B infection and for immune serum globulin prophylaxis of dental personnel with "needle sticks" or other exposure to HBsAg positive patients.

The frequency with which hepatitis B virus is transmitted to patients in dental offices is probably quite low, and far less than the frequency with which it is transmitted from patients to dental personnel in the same setting (8,9). Several instances in which hepatitis B infection was transmitted to a fairly large number of patients in the same dental practice have been studied and reported in national journals (3,4), but these are apparently isolated instances, and the USPHS Centers for Disease Control cited only a total of eight such incidents in a recent publication (9). In most cases, transmission of hepatitis B to dental patients probably occurs from wounds or eczematous lesions on an infected dentist's hands, and inadequately sterilized instruments or local anesthetic needles are not thought to be involved. Epidemics of hepatitis B infection have in the past been associated with contaminated needles, syringes and multidose vials, but hazards of this kind became almost nonexistent in dental practice in the 1960's with the introduction of disposable needles and single dose carpules of local anesthetic as a standard procedure in dentistry (10,11). Surveys of infections with hepatitis B and hepatitis A carried out in recent years (9) have also demonstrated that dental visits are no more common as antecedent events for hepatitis B infections than they are for hepatitis A infections, which are not commonly considered to be syringe-borne infections (12).

METHODS SUGGESTED FOR CONTROL OF HEPATITIS B INFECTION OF DENTAL PERSONNEL

Concern about the possibility of acquiring chronic infection with hepatitis B has resulted in a number of procedures being recommended for control of transmission of the virus to dental personnel. These include 1) use of barrier protection methods; 2) screening of dental patients for HBsAg carrier status and establishing special treatment facilities for known carriers; and 3) use of passive and active immunization.

1. Use of barrier protection methods. Routine use of gloves by dentists has been recommended as a means of reducing the opportunity for transmission of hepatitis B from infected patient to dentist (13), although the effectiveness of gloves in this context has not been documented. The protective value of gloves is compromised by the frequency with which they are punctured or torn by sharp cavity margins, rough bone spicules or sharp dental equipment items such as burrs, matrix bands, wire ligatures, etc. Use of gloves by oral surgeons and periodontists is relatively common, but may be more effective in preventing transmission of hepatitis B to patients from a carrier dentist than transmission from patient to dentist. In the eight episodes of suspected transmission of hepatitis B infection to dental patients studied by the USPHS Centers for Disease Control (9), none of the dentists implicated routinely wore gloves, and during a subsequent period of observation when they were obliged to wear gloves, no additional transmission of hepatitis B was noted. The tendency for traces of blood to remain around the finger nails and in the creases of the fingers and hands despite hand washing has also been documented (14) and may partially explain the effectiveness of gloves in preventing transmission of hepatitis B infections to patients. Hepatitis B virus has been transmitted by aerosolized suspensions in experimental animal studies (15), suggesting that the wearing of a mask and eye glasses may provide additional protection for dental personnel exposed to aerosols of saliva from HBsAg positive carriers (6).

2. Screening of dental patients for HBsAg carrier status. Health questionnaires commonly used in dental offices (11) include questions regarding prior episodes of hepatitis, jaundice, rejection as a blood donor, and intravenous drug abuse, that are designed to identify the patient who may be at an increased risk for transmission of hepatitis B infection. Since the majority of hepatitis B infections are subclinical, the effectiveness of this approach has been questioned, and one study of 140 patients registering in a dental school clinic (16) showed that only 50% of those subsequently found to be HBsAg positive gave a history of hepatitis. The effectiveness of screening for HBsAg carriers can presumably be increased by also testing those in high-risk groups

for acquiring hepatitis B infection. As currently stated, these groups are: hemodialysis patients and patients who have received a renal transplant; individuals who have been institutionalized for mental treatment or penal service for over six months; narcotic addicts and those on methadone or similar replacement therapy; military personnel or others who have lived "off base" for longer than six months in Mediterranean, tropical, S.E. Asian and other countries where the prevalence of hepatitis B is known to be high; migrants from these countries; and individuals reporting multiple sexual contacts (whether of a heterosexual or homosexual nature), in addition to those who give a history of hepatitis or jaundice. Results of such a screening program in a dental school clinic population in which prospective patients in these high-risk groups and those with a history of hepatitis were required to be screened for HBsAg are illustrated in Table 1.

While this approach is intended to identify the hepatitis B carriers so that special attention can be given to control of the hazards of infectious transmission associated with dental care, it has the disadvantage that individuals identified as HBsAg carriers are often stigmatized on this account and subsequently find it very difficult to obtain dental care unless they hide the fact that they are hepatitis B carriers (18).

Table 1. Hepatitis Bs antigenemia and history of hepatitis in 7600 patients registering at the University of Pennsylvania School of Dental Medicine, September 1979-March 1981.

	Percentage of Patients Registering	Percentage of Patients in Group
Patients with a history of hepatitis	137/7600 (1.7%)	4/137 (3%)
S.E. Asian immigrants	40/7600 (0.5%)	1/40 (2.5%)
Total # patients screened for HBsAg	177/7600 (2%)	5/177 (3%)
Total # patients registered	7600 (100%)	5/7600 (0.07%)[a]

[a]Assuming "true frequency of HBs antigenemia" is approximately 5x frequency with which HBsAg is detected among those with a history of hepatitis (Tullman, 1980, Ref. 17), prevalence of HBs antigenemia in this population sample is 0.35%, or 15-20 cases per year.

3. Use of passive and active immunization procedures. Passive immunization for hepatitis B infection has been available for a number of years in the form of immune serum globulin (ISG)* or hyperimmune hepatitis B globulin (HBIG).** The latter has been used for prophylaxis of contacts of patients with acute hepatitis B infection (e.g. spouses of these patients, and institutional or family contacts of retarded children with acute hepatitis B) and for management of clinical personnel with known recent percutaneous contact with this infection (e.g. accidental needle sticks or surgical lacerations). The limited life of this passively acquired immunity and the need for the globulin to be administered within three days of the contact seriously reduce the value of this approach, however.

While the possibility of active immunization against hepatitis B infection was demonstrated over ten years ago by Krugman's finding that heat-inactivated hepatitis B serum is noninfectious but antigenic and protective against subsequent challenge with hepatitis B infection in institutionalized children (19), it was only recently that purified preparations of HBsAg free of contaminating intact infectious virus have been available for clinical trials (20,21). Such preparations first became available for use on a group of dental personnel in 1981 when Merck Sharp and Dohme purified HBsAg vaccine was administered to some 110 faculty, staff and students of the University of Pennsylvania School of Dental Medicine as part of the pre-licensing batch consistency trials required of the product that has been subsequently marketed as Heptavax B.

TRIAL OF HEPATITIS B VACCINE IN DENTAL PERSONNEL

Volunteers for this trial were solicited from among the then current faculty, staff and students of the University of Pennsylvania School of Dental Medicine. Potential participants were informed that they would be required to have screening blood tests for HBsAg, anti-HBs and anti-HBc antibody, and SGOT and SGPT enzyme levels. Individuals who were found to be without evidence of prior contact with hepatitis B and who had levels of SGOT and SGPT within the normal range were given the opportunity to receive three doses of the experimental vaccine (in open fashion no control group would be used) under the conditions proposed for clinical use of the vaccine, *viz.* three doses administered intramuscularly

---

*Immune serum globulin (human) USP.

**HeP-B-gammagee, Merck Sharp and Dohme, Division of Merck and Company, Inc., West Point, PA 19486.

at times zero, one month and six months, respectively. Participants were also required to have additional blood specimens drawn at three, six and seven months after the initial dose for repeated measurement of antibody response. Dental personnel participating in this trial were one segment of a total of 395 health care personnel at the University of Pennsylvania and The Children's Hospital of Philadelphia included in the testing of several batches of the vaccine.

A total of 133 dental personnel (56 staff and faculty, and 77 students) were screened for the purposes of the trial (see Table 2). Of those screened, 13 or 9.8% (16.0% for staff and faculty, 5.2% for students) were found to have anti-HBs or HBc antibodies, and HBsAg was detected in one individual; no elevated SGOT or SGPT levels were detected. The 14 individuals with evidence of prior infection with the virus were excluded from the study. Another 10 dropped out before completion of the study, leaving a total of 109 (42 staff and faculty, and 67 students) who received three doses of vaccine and on whom follow-up antibody data through the seven-month period following the initial dose are available.

The vaccine* (a formalin inactivated, alum-absorbed subparticulate hepatitis B vaccine prepared to contain 20 μg/ml of HBsAg protein, 500 μg/ml of alum and 1:20,000 thimerosal preservative) was administered intramuscularly in three doses. Forty-eight of the 109 received 1.0 ml (20 μg of HBsAg protein), and the other 61 received 0.5 ml (10 μg of HBsAg protein) doses. All serum samples ("zero," i.e. one week before the first dose, three months, six months and seven months) were assayed by Merck Sharp and Dohme Research Laboratories as follows: HBsAg (AUSRIA-II),** anti-HBs (AUSAB),** anti-HBc (CORAB),** SGOT and SGPT (standard clinical chemistry laboratory assays). Serum containing antibody to hepatitis B in titers of $<$ 1:8 were considered to be negative, as were those with SGOT and SGPT levels less than 40 units/ml. Seroconversion rates were calculated for those individuals remaining in the study at each of the three specimen collection periods and expressed as: the ratio (%) of number with anti-HBs/number vaccinated; geometric mean titers of anti-HBs were calculated according to the directions supplied by the manufacturer of AUSAB. All vaccinated individuals maintained a written record of oral temperature and any local or systemic complaints for five days following each dose.

---

*Heptavax B, Merck Sharp and Dohme Research Laboratories, West Point, PA 19486.

**Abbott Laboratories, North Chicago, IL 60064.

Table 2. Participation of faculty, staff and students of the University of Pennsylvania School of Dental Medicine in a trial of Merck Sharp and Dohme hepatitis B vaccine, January-December 1981.

	Faculty and Staff	Students	Total
# screened for hepatitis B markers	56	77	133
# with antibody to hepatitis B	9 (9/56-16.0%)	4 (4/77-5.2%)	13 (13/133-9.8%)
# with HBsAg	0	1	1
# lost from study for other reasons	5	5	10
# remaining in trial for a full seven-month period	42	67	109

The results of this trial are illustrated in Tables 3, 4 and 5. Overall response to the vaccine was good with over 95% of all participants developing anti-HBs titers one month after the third dose. This response rate is similar to that reported in other trials of Heptavax B (22). There were no significant differences in the responses of persons receiving three doses of 10 µg HBsAg (one half the currently recommended dose of Heptavax B) versus three doses of 20 µg HBsAg. There were also no differences in the responses of males versus females. However, serum conversion rates and geometric mean titers were significantly lower at all times among vaccinees 40 years of age or older as opposed to those less than 40 years of age. This difference has been noted in other trials of Heptavax B (23). Clinical complaints including temperature elevations to 99°F were minimal, well tolerated and unrelated to the particular batch of vaccine.

POSTSCRIPT

Following licensing of Heptavax B in November 1981, a questionnaire was distributed to 435 faculty and 428 students at the University of Pennsylvania School of Dental Medicine inquiring if they wished to be immunized with this new vaccine. Responses were received from 30% of the faculty and 22% of the students. Of those

Table 3. Trial of Merck Sharp and Dohme hepatitis B vaccine in dental personnel, University of Pennsylvania School of Dental Medicine, January-December 1981.

Comparison of seroconversion rates and geometric mean titers (GMT) by Dose[a]

Time	Dose[b] = 10 ug			Dose[b] = 20 ug		
	# with anti-HBs / # vaccinated (%)	GMT All	GMT Responders only	# with anti-HBs / # vaccinated (%)	GMT All	GMT Responders only
3 Mos.	56/62 (90.3)	158.2	272.2	44/49 (89.8)	173.4	311.6
6 Mos.	56/61 (91.8)	251.9	412.7	46/49 (93.9)	358.5	526.1
7 Mos.	58/61 (95.1)	7340.5	11,632.8	46/48 (95.8)	4511.7	6504.8

[a]Data from Brightman, V. J., et al., Immunization of dental personnel with inactivated hepatitis B vaccine. Proc. Am. Acad. Oral Path., Ann. Mtg., Reno, NV, May 1982.
[b]ug HBsAg protein/ml.

Table 4. Trial of Merck Sharp and Dohme hepatitis B vaccine in dental personnel, University of Pennsylvania School of Dental Medicine, January-December 1981.

Comparison of seroconversion rates and geometric mean titers (GMT) by Sex[a]

Time	Sex = Male			Sex = Female		
	# with anti-HBs / # vaccinated (%)	GMT All	GMT Responders only	# with anti-HBs / # vaccinated (%)	GMT All	GMT Responders only
3 Mos.	71/79 (89.9)	165.4	294.1	29/32 (90.6)	163.1	276.5
6 Mos.	72/78 (92.3)	302.8	427.0	30/32 (93.8)	371.6	551.4
7 Mos.	73/77 (94.8)	5311.0	8497.6	31/32 (96.9)	7707.0	10,285.9

[a]Data from Brightman, V. J., et al., Immunization of dental personnel with inactivated hepatitis B vaccine. Proc. Am. Acad. Oral Path., Ann. Mtg., Reno, NV, May 1982 (modified from reference 1, Table 28-7, courtesy J. B. Lippincott Co., Philadelphia, PA).

Table 5. Trial of Merck Sharp and Dohme hepatitis B vaccine in dental personnel, University of Pennsylvania School of Dental Medicine, January-December 1981.

Comparison of seroconversion rates and geometric mean titers (GMT) by age range[a]

	Age < 40 years			Age ≥ 40 years		
	# with anti-HBs / # vaccinated (%)	GMT		# with anti-HBs / # vaccinated (%)	GMT	
Time		All	Responders only		All	Responders only
3 Mos.	85/90 (94.4)	275.9	354.2	15/21 (71.4)	25.1	91.1
6 Mos.	86/89 (96.6)	461.4	571.5	16/21 (76.2)	44.1	144.1
7 Mos.	86/88 (97.7)	11,042.4	13,711.6	18/21 (85.7)	435.9	1200.6

[a]Data from Brightman, V. J., et al., Immunization of dental personnel with inactivated hepatitis B vaccine. Proc. Am. Acad. Oral Path., Ann. Mtg., Reno, NV, May 1982.

responding, only 52 students (12% of the student body) and 61 faculty (14% of the total faculty) indicated their willingness to be vaccinated. This lack of interest in the vaccine was surprising considering the concern that continued to be expressed about the risks of acquiring hepatitis B infection in clinical dental practice, the fact that dental personnel were listed as a high priority group for receiving the vaccine (24), and endorsement of the vaccine by the Council on Dental Therapeutics of the American Dental Association (25) and component state dental societies (26,27). Cost of the vaccine was given as a reason for delaying vaccination by only a minority of the respondents. Many preferred to wait until there had been more experience with the vaccine, and they expressed concern about the risk of acquiring hepatitis B or other infections such as AIDS from the vaccine. Consistent with these questionnaire responses, very few faculty and students utilized immunization programs that were established for this purpose in the Student Health Service and the hospital Occupational Health Clinic, and a second survey carried out at the School of Dental Medicine in May 1983 revealed that only 13% of the faculty and 14-28% of the students with clinical contact were known to be vaccinated or naturally immune to hepatitis B infection. Additional efforts were then made to publicize the need for hepatitis B vaccination and the safety of Heptavax B among the school's faculty and students. [The Council of Dental Therapeutics of the American Dental Association like a number of other national review panels also issued statements on the safety of Heptavax B at this time (28)]. In recent months (January, February 1984) we have noted an increased interest in receiving the vaccine among both faculty and students, and as many as 20% of the student body have reported for vaccination when the vaccine was administered at the school itself. Initial concern about the safety of the vaccine appears to be abating, and the possibility of a dental institution in which the majority of the faculty, staff and students with clinic assignments are immune to hepatitis B infection now seems achievable.

ACKNOWLEDGMENTS

Supported by National Institutes of Health General Clinical Research Centers, Division of Research Resources #5-M01-RR01224-04 and Merck Sharp and Dohme Research Laboratories.

REFERENCES

1. Brightman, V. J. (1983). Sexually transmitted diseases, _in_ Burket's Oral Medicine, 8th edition, M. A. Lynch et al., eds., J. B. Lippincott, Philadelphia, Chapter 28.

2. Smith, J. L., et al. (1976). Comparative risk of hepatitis B among physicians and dentists. J. Infect. Dis.133:705.
3. Rimland, D., et al. (1977). Hepatitis outbreak traced to an oral surgeon. N. Engl. J. Med. 296:953.
4. Hadler, S. C., et al. (1981). An outbreak of hepatitis B in a dental practice. Ann. Intern. Med. 95:133.
5. USPHS Centers for Disease Control. (1980). Viral hepatitis B, tuberculosis and dental care of Indochinese refugees. Morb. Mort. Wkly. Rpts. 29:1.
6. Council on Dental Therapeutics, Am. Dental Assoc. (1976). Type B (serum) hepatitis and dental practice. J. Am. Dent. Assoc. 92:153.
7. Expert Group on Hepatitis in Dentistry. (1979). Hepatitis in dentistry. Brit. Dent. J. 146:123.
8. Tzukert, A. and Gandler, S. G. (1978). Dental care and spread of hepatitis B infection. J. Clin. Microbiol. 8:302.
9. Ahtone, J. and Goodman, R. A. (1983). Hepatitis B and dental personnel: transmission to patients and prevention issues. J. Am. Dent. Assoc. 106:219.
10. Council on Dental Materials and Devices, Council on Dental Therapeutics, Am. Dental Assoc. (1978). Infection control in the dental office. J. Am. Dent. Assoc. 97:673.
11. Am. Dental Assoc. (1979). Accepted Dental Therapeutics, 38th Edition, Am. Dental Assoc., Chicago, IL.
12. Mosley, J. W. (1978). Epidemiology of HAV infection, _in_ Viral Hepatitis, A Contemporary Assessment of Etiology, Epidemiology, Pathogenesis and Prevention, G. N. Vyas, et al., eds., Franklin Institute Press, Philadelphia, Chapter 8.
13. Allard, W. F. (1984). The significance of rubber gloves: reduction of transmission of infectious diseases in dental practice. Penna. Dent. J. Jan.-Feb., p. 20.
14. Allen, A. L. and Organ, R. J. (1982). Occult blood accumulation under the fingernails: a mechanism for the spread of blood-borne infection. J. Am. Dent. Assoc. 105:455.
15. Bancroft, W. H., et al. (1977). Transmission of hepatitis B virus to gibbons by exposure to human saliva containing hepatitis B surface antigen. J. Infect. Dis. 135:79.
16. Goebel, W. M. (1979). Reliability of the medical history in identifying patients likely to place dentists at an increased hepatitis risk. J. Am. Dent. Assoc. 98:907.
17. Tullman, M., et al. (1980). The threat of hepatitis B from dental school patients. A one-year study. J. Oral Surg. 49:214.
18. Editorial. (1975). The HBV carrier - a new kind of leper. N. Engl. J. Med. 292:477.

19. Krugman, S., et al. (1971). Viral hepatitis, type B (MS-2 strain). Studies on active immunization. JAMA 217:41.
20. Hilleman, M. R., et al. (1978). Clinical and laboratory studies of HBsAg vaccine, _in_ Viral Hepatitis, A Contemporary Assessment of Etiology, Epidemiology, Pathogenesis and Prevention, G. N. Vyas, et al., eds., Franklin Institute Press, Philadelphia, Chapter 51.
21. Purcell, R. H. and Gerin, J. H. (1978). Hepatitis B vaccines: A status report, _in_ G. N. Vyas, et al., eds., op. Cit. Chapter 49.
22. Szmuness, W., et al. (1981). A controlled clinical trial of the efficacy of the hepatitis B vaccine (HEPTAVAX-B): a final report. Hepatology 1:377.
23. Hilleman, M. R. Personal communication.
24. Krugman, S. (1982). The newly licensed hepatitis B vaccine, characteristics and indications for use. JAMA 247:2012.
25. Council on Dental Therapeutics, Am. Dental Assoc. (1982). ADA Council recommends hepatitis vaccines for dentists, students, auxiliary personnel. ADA News, Aug. 16, p. 1.
26. Pennsylvania Dental Assoc. (1983). Hepatitis B - a high risk to dental personnel. Penna. Dent. J. 50:47.
27. New Jersey Dental Assoc. (1982). NJDA offers Heptavax B vaccine. ADA News, Nov. 22, p. 1.
28. Council on Dental Therapeutics, Am. Dental Assoc. (1983). Hepatitis B vaccine safety attested. Council finds no evidence of AIDS risk. ADA News, Aug. 29, p. 1.

THE DEVELOPMENT OF THE HEPATITIS B VACCINE

Irving Millman

Institute for Cancer Research
Fox Chase Cancer Center
Philadelphia, Pennsylvania 19111

In the October 9, 1980 issue of The New England Journal of Medicine Szmuness et al. published the results of a clinical trial. This trial assessed the efficacy of an inactivated hepatitis B vaccine (produced by Merck) in a placebo controlled, randomized, double blind trial in 1083 homosexual men at high risk for hepatitis B virus infection. Within the limits of the studies that have been reported, the vaccine was found to be safe; the incidence of side effects was low. After a third dose was administered, 96% of those who received the vaccine developed antibodies against HBsAg and none of the individuals who developed antibodies had a hepatitis episode. The results of this trial were discussed at the 1981 International Symposium on Viral Hepatitis in New York City March 30 through April 2, 1981, and celebrations appeared to be in order for the first hepatitis B vaccine to be placed on the market.

In January 1982 we learned that the Institut Pasteur had marketed a hepatitis B vaccine called "Hevac B Pasteur." The Merck vaccine did not appear on the market until June or July 1982. Both vaccines are similar and differ only in production procedures which have never been fully disclosed. This paper will give an account of the role of the Institute for Cancer Research in the discovery and development of the vaccine.

When I was first interviewed in May 1967 for a position at the Institute for Cancer Research, I thought that I would be working on a blood substance which was of interest because it appeared to be polymorphic. On May 1, 1967 I watched Barbara Werner set up a density gradient of fractions where this substance was concentrated, and when I returned to the Institute in July I was surprised to learn that Barbara was in the hospital with hepatitis B. She actually diagnosed her own disease by finding the hepatitis surface antigen, then called Australia antigen, in her blood. The association between hepatitis B and the Australia antigen had been suggested in November 1966 (1). However, the full impact of the connection registered with me only then. My thoughts of leaving the field of infectious agents for one of pure immunology faded fast. Work had to be delayed until needed equipment such as laminar flow hoods could be installed. More than anything else we had to change our thinking concerning the handling and storage of thousands of blood samples and concentrated serum fractions that most likely could transmit hepatitis.

The fractions that we were dealing with contained particles which looked like viruses, as published in _Nature_ in June 1968 (2). In this paper it was stated that "Australia antigen found in patients with the diagnosis of leukemia, Down's syndrome and hepatitis is associated with particles, and it is important to determine their origin and function." Although these particles looked like viruses, I could not originally accept that idea. For one thing, we estimated that many of the sera contained as much as 10^{13} particles per ml. By comparison, a concentration of bacteriophage equivalent to this would be a gel. Our most purified fractions did not contain nucleic acids as would be expected if they contained intact infectious viruses and, I reasoned, furthermore, that no animal could tolerate this much virus and live. We came to the conclusion that we were either dealing with incomplete viruses or parts of a virus. Our earlier attempts at replicating the hepatitis B virus in tissue or organ cultures appeared promising in that we could show by fluorescent antibody techniques the presence of surface antigen in 2 liver cultures out of 23 specimens from patients with surface antigen in their blood. This was shown in various passages of these cultures and also confirmed by radioimmunoprecipitation assay of tissue culture fluids (3). However, these experiments were difficult to repeat and were not confirmed by others. We could not produce growth of hepatitis B virus by conventional methods in normal human tissue or organ cultures.

During 1968 and 1969 we and others began accumulating evidence that antibody to the surface antigen might be protective. Okochi and his colleagues in Japan (4,5) reported that blood transfusion recipients with antibody to the surface antigen were less susceptible to hepatitis B infection. Okochi reported that, while only 6 of 10 multiply transfused patients who developed anti-HBs also developed signs of hepatitis, 42 of 45 who did not develop anti-HBs did develop hepatitis. This difference was significant with a probability of < 0.015.

Our early studies of HBsAg in animals were designed to determine the distribution of this antigen in nature and also to look for a suitable animal for assays of infectivity. Our findings (6) and those by Lichter (7) and Hirschman et al. (8) verified that only certain non-human primates had HBsAg. In addition to chimpanzees, we found antigen in vervets (Cero-pethecus pygerythrus), marmosets and red spider monkeys. Hirschman detected the antigen in gibbons and orangutans. To determine infectivity of serum containing HBsAg we inoculated 6 weanling vervets with whole human serum containing this antigen. None of these developed detectable antigen, although one developed antibody after a single intravenous injection. Next, we inoculated 2 adult vervets with serum containing HBsAg and again neither of these animals developed HBsAg in their blood. We then attempted to learn what happened to HBsAg after it was injected. HBsAg was partially purified. One ml of the fractions with the highest immunoreactivity was diluted and injected in divided doses intramuscularly, intravenously, and intraperitoneally into a 7-day-old African green monkey. The monkey was sacrificed 24 hours after inoculation. As estimated by immunodiffusion titration, the total blood volume of the monkey contained more HBsAg than was inoculated. Based on this observation, an experiment was devised to determine whether there was replication in infant vervets and whether a virus could be passaged. In this experiment 100 ml of plasma containing HBsAg was pelleted and the pellet was treated with a series of degradative enzymes for proteins, lipids, carbohydrates and nucleic acids. This material was then passed through a Sephadex G200 column and the first peak was recentrifuged to pellet the immunoreactive material. This resulting fraction was resuspended in Hank's balanced salt solution, diluted, and 1.5 ml portions were inoculated IM, IV and IP into a 7-day-old male African green monkey that weighed 400 g. The first recipient was sacrificed in 24 hours and 1 ml of his serum was diluted and inoculated into a second baby monkey. Serum from the second monkey was subsequently administered to a third monkey. The serum of the third monkey became positive 6 months after inoculation. These experiments demonstrated transmission of partially purified HBsAg fraction to a non-human host, replication of HBsAg in that host and two passages of HBV (9,10,11).

Next there followed a series of experiments (unpublished) where highly purified fractions of hepatitis surface antigen were used in a similar fashion. In none of the animals injected were there signs of replication. Only the appearance of anti-HBs gave evidence that the animals had been injected with a preparation of hepatitis surface antigen.

At this point in early 1969, we felt that the surface antigen which we called Australia antigen (Au(1)) was a surface component of the HBV. Antibody to only a surface component could be protective because it seemed reasonable to assume that the attachment to susceptible hepatocytes would be through surface antigen receptors and that protective antibody would have to operate by blocking the HBV from attachment to the hepatocyte.

Now at this point I would like to digress for a few moments and recall the times. During the Johnson-Nixon years, the ability to obtain sufficient funding for research was a problem. To alleviate this problem, the United States Government conceived the "patent policy" which meant that a non-profit organization such as a university, hospital or research institute could assume ownership of a patent which could then be sold. This was a sound idea because it allowed a product to be sold to industry where it could be developed and marketed. Funds from a sale could go back into research. This apparent gift of the government was a two-way affair, however, since the idea was that less funding would be required by research groups and that the private sector would more or less take over this function.

Now, to recapitulate, we had evidence that:

1. HBsAg was the surface component of HBV.

2. Antibody to HBsAg might be protective against infection.

3. We could purify HBsAg free of nucleic acids and serum proteins.

4. There was a ready supply of surface antigen in human carrier blood. Hence, it was not necessary to grow hepatitis B virus in culture or by other means in order to obtain large quantities of surface antigen.

5. The government had requested that basic research should be applied and made available for general medical use.

In October 1969, acting on behalf of the Institute for Cancer Research, Dr. Blumberg and I filed an application for a patent for a hepatitis B surface antigen vaccine and the process for its production. This patent was subsequently (January 1972) granted in the United States and several other countries. The abstract of the patent disclosure is as follows: "A vaccine against viral hepatitis is derived from blood containing Australia antigen, having particles resembling viruses which are substantially free of nucleic acid, of a size range of 180-210 Å, substantially free of infectious particles. The vaccine where required is attenuated or altered. The preferred procedure for removing impurities including infectious components involves centrifugation, enzyme digestion, column gel filtration, differential density centrifugation in a solution of sucrose, dialysis, differential density centrifugation in a solution of cesium chloride and dialysis."

Two events occurred at about this time which gave us encouragement. Krugman's somewhat controversial reports regarding the effects of the administration of a pool of serum contaminated with hepatitis virus on mentally retarded patients was making headlines. His material collected from one patient with two episodes of jaundice (MS-1 and MS-2) (12) provided evidence that there were actually two viruses, a hepatitis A and a hepatitis B. His attempts to produce a vaccine using heated whole sera was the controversial part. His interpretation was that heated sera (which contained HBsAg) could be administered to induce protective antibody. Under his experimental conditions it was difficult to determine whether the children reacted to what was given to them experimentally or to a natural infection. In other experiments described by Krugman (13) he challenged the children with material containing hepatitis virus and several came down with typical viral hepatitis.

Our Clinical Research Unit at the Institute decided that we would not do the field studies required to test the vaccine we had proposed. Trials of this nature would have required an expenditure in time and money which we felt should be expended on our other commitments in research. Also, we decided it would be better to have someone other than ourselves test the efficacy of the vaccine we had introduced. Furthermore, it was questionable whether the Bureau of Biologics, Standards, the current FDA, would approve a vaccine which came from human blood or contained even traces of human blood. The "minimum requirements" set forth by the Federal Security Agency and regulated by the Bureau of Biologics, Standards forbade the presence of blood or traces of blood in manufactured vaccines such as DPT (diphtheria, pertussis, tetanus). We decided to contact the pharmaceutical house of Merck and Company, which was located near the Institute, which we felt had the expertise for such a task. On September 14, 1971 we received a written commitment from Dr. Maurice Hilleman of Merck and Company to undertake the development of the vaccine and expend 5 million dollars on a project involving 7-10 technical personnel to this end. He stated that he hoped to have a vaccine on the market within five years. For such an expenditure Merck and Company wanted exclusive rights to the patent. We learned shortly thereafter that we could not give this because our Institute had neglected to apply for an institutional patent agreement. Only the National Cancer Institute (NCI) of the NIH agreed to allow our Institute to apply for an institutional patent agreement and assume ownership of the patent; therefore, it became NIH property. The NIH proposed to license any or all approved companies which paid a licensing fee to manufacture the vaccine as taught in the patent. As expected, Merck and Company did not appear interested and direct contact between our Institute and this company did not re-occur for five years. After much discussion with the NIH, a compromise was reached and the Institute for Cancer Research was granted foreign but not domestic rights to the patent.

Prior to our negotiations with the NIH for patent ownership, interest in the vaccine became more general. Krugman (13) published his experiments where MS-2, diluted with distilled water and heated to 98°C for 1 minute, was injected into 29 susceptible children from Willowbrook State School, an institution for the mentally retarded located in New York City. About half of the children received three inoculations of the inactivated vaccine. Four to eight months after the last inoculation they were challenged with unheated serum containing active virus. Krugman claimed that in 20 (69%) of the 29 susceptible children type B hepatitis was prevented or modified. The disease was attenuated in the remaining children and was associated with a reduction in the incidence of a chronic carrier state. In June 1971 Soulier, at a seminar held at the International Children's Center in Paris, reported that heated serum containing hepatitis B virus could be used to protect against infection. Soulier heated his preparations for 10 hours at 60°C (14).

Shortly thereafter systematic studies using experimental animals were begun. Purcell and Gerin (15) at the National Institutes of Health began evaluating their version of a purified vaccine which was produced generally in a similar manner to the one we proposed and patented. One unit of asymptomatic carrier blood of subtypes ayw and adr were each purified and made up as vaccine. This vaccine consisted primarily of 22 nm particles purified by isopycnic banding in cesium chloride and rate sedimentation centrifugation. It was inactivated by incubation in a 1:2000 formalin solution for 96 hours at 37°C. The concentration of protein was adjusted to 20 μg/ml (Lowry). This vaccine required stabilization with 1.4% human serum albumin. Purcell and Gerin used an immunizing dose of 1.0 ml subcutaneously followed by a "booster" injection of the same dose subcutaneously a month later. Seven chimpanzees were used in the tests. Sera from the 7 chimpanzees were tested for SGPT, HBsAg, anti-HBc, and anti-HBs weekly for 24 weeks. Only the chimpanzee inoculated with the HBsAg ayw positive serum became infected and developed elevated SGPT, anti-HBc and HBsAg. The other 6 animals were not apparently infected and showed none of these changes. The vaccinated animals were challenged with HBV (MS-2 strain, ayw) in an amount calculated to cause hepatitis in 50% of the chimpanzees 24 weeks after the final vaccination injection. Blood samples were collected twice a week after challenge. Both control animals (one which received no injection and one which received the diluent) developed HBs antigenemia and anti-HBc. In contrast, none of the immunized animals developed detectable HBsAg in their sera nor did they develop evidence of hepatitis.

In February 1975 Dr. Blumberg and I embarked for England and France in an effort to find a manufacturer for the vaccine and get the vaccine out to market with the least delay. I felt that efforts of the NIH toward developing a vaccine were non-productive since they could not market it themselves. The NIH could only license; and no company, in my opinion, appeared willing to expend much money without the guarantee of an exclusive concerning the patent. Our first meeting was with the Wellcome Research Laboratories. It became evident that they were not interested and, according to Blumberg's recollection, "They didn't even invite us to tea." Our next meeting was with the Glaxo people in Liverpool. After a delightful welcome and frank discussions they indicated that they were "vitally" interested in the vaccine and that they would contact us after they had investigated the commercial possibilities. I. B. Smith, Managing Director of Duncan Flockhart, Ltd. (the marketing section of Glaxo holdings), stated that Glaxo had a special concern with public health problems and that they wanted to prevent disease. We left with the assurance of a major interest and that they would contact us within a few weeks with an offer. Our final meeting was in Paris with the Institut Merieux people. Robert Lang, Director of Research, was very interested and discussions revolved around marketing and patent rights. Charles Merieux (the son of the founder) also participated in these discussions. We left for home optimistic about the possibility of finally getting the vaccine developed for trials and the market. We did not hear from our foreign contacts immediately. In early 1975 I again approached (this time informally) the Merck people at a hepatitis conference held in Bethesda. By this time, the Institute had obtained foreign patents in several countries in Europe and Asia. Merck and Company expressed interest in the acquisition of our foreign patent rights. Mr. G. Willing Pepper, who was Chairman of the Institute's Board of Trustees, was instrumental in starting discussions with Merck's top management. While these negotiations were going on, the National Information Service of the United States Department of Commerce contacted the Institute with a proposal. The Institute for Cancer Research could act as agent for the Department of Commerce in setting up non-exclusive licensing for development of the vaccine. This non-exclusive licensing would include the domestic as well as all of the Institute's foreign patent rights. The financial aspects of such an agreement would be negotiable. Regarding foreign patents, an agreement had been reached earlier between the Institute and the Secretary of HEW. The U.S. patent would remain the property of the government while all foreign patents would remain the property of the Institute for Cancer Research. In August 1975 a decision was reached to go along with

Merck and Company. An agreement was drawn up whereby Merck would obtain all foreign patent rights regarding the hepatitis B vaccine. Merck would obtain the necessary license from the United States Food and Drug Administration for domestic rights on an apparently non-exclusive basis. Shortly thereafter Merck and Cutter laboratories were licensed by the FDA to produce the vaccine.

Events then proceeded rapidly and effectively under the supervision of Dr. Maurice Hilleman, the person at Merck responsible for vaccine development and production. In March 1975 and in June 1976 Merck reported on their chimpanzee protection studies (16,17). They found that 3 doses of the vaccine protected chimpanzees against 1000 chimpanzee infectious doses of live hepatitis B virus given i.v. Six chimpanzees were vaccinated before challenge and there were 5 unvaccinated controls. The animals were observed for 24 weeks. None of the vaccinated animals developed clinical evidence of hepatitis.

Antibody to the group subtype "a" appeared to be sufficient for protection. Murphy, Maynard and LeBouvier (18) reported that in experiments where chimpanzees were inoculated with various unpurified subtype-specific preparations of HBsAg the animals became infected and developed HBsAg and eventually anti-HBs that corresponded with the original inoculum. An extension of these studies by Barker, Gerety et al. (19) indicated, however, that convalescent chimpanzees were protected from infection with heterologous subtype hepatitis B virus preparations. Either mixtures of the various subtypes would be made or vaccines would be composed of subtypes found prevalent in an area where the vaccine would be used.

Our French colleague, Dr. Philippe Maupas, spent several months in our laboratory beginning in October 1974 working on various problems related to hepatitis virus. Later he and his colleagues (20) in Tours reported on their experience with an HBsAg vaccine. They published data on a vaccine which was purified by affinity chromatography using anti-HBs coupled by cyanogen bromide to Sepharose 4B. The purified material, standardized to 70 μg/ml protein still contained traces of IgG. The vaccine was inactivated with 0.1% formalin for 48 hours at 37°C and one week at 4°C. It was found safe when administered to 5 chimpanzees, i.e. none of the animals had evidence of infection with hepatitis B virus. Maupas next vaccinated 30 patients and 73 staff members of a renal dialysis unit in Tours. Each vaccinated individual received 2 subcutaneous injections of 1.0 ml of vaccine within a month. He reported that 95% of the staff and 60% of the patients developed anti-HBs. Both staff and patients were followed for a year. None of the vaccinated group developed clinical hepatitis or became HBsAg positive. Fifty percent of the unvaccinated staff and 60% of unvaccinated patients became

HBsAg positive. Included among the apparently protected staff were individuals who did not develop anti-HBs following vaccination. On the basis of this observation, Maupas et al. suggested that protection may develop even in the absence of detectable anti-HBs in the blood. He also postulated that cellular immunity might play a protective role. Maupas was the first to use an aluminum hydroxide adjuvant with the vaccine; the use of adjuvant resulted in a more rapid seroconversion.

Brzosko et al. in "Hepatitis Scientific Memoranda" for March 1976 reported preliminary results with a similar purified vaccine. Their first lot of vaccine was purified by affinity chromatography, standardized to 200 μg/ml of protein and inactivated with 1:100 formalin for 168 hours (the temperature of inactivation was not given). The formalin was removed by dialysis. Three chimpanzees were inoculated with massive doses of the vaccine to check its safety. Each animal received 10 ml of vaccine for a total of 2 mg of HBsAg, by the intramuscular (i.m.) and i.v. route. Nine weeks later the animals were given i.m. booster injections of 5 ml of vaccine. Two months after the first booster the animals were given an additional 0.1 ml of vaccine i.m. Four weeks after this last injection the animals were challenged with HBV.

Because chimpanzees are expensive and difficult to obtain, the same animals were used for immunogenicity control studies of a second lot of vaccine ten weeks later. This lot was prepared from serum containing HBsAg by precipitation with ammonium sulfate, followed by digestion with pepsin and concentration on an Amicon DC 30 equipped with XM 100 filters. They claimed that HBsAg was detected in each animal only once, in the serum sample collected one week after the first vaccine injection. Anti-HBc was never found in any of the animals and anti-HBs was detected in all animals after the first week. They concluded that HBsAg vaccines, made by their procedures, were safe.

Because of the theoretical possibility that the vaccine might contain host nucleic acid and/or human, cellular, or serum antigens, polypeptide vaccines have been proposed. Purified HBsAg would be degraded to polypeptides and each separated in the expectation that the anti-HBs inducing host characteristics would be found in different fractions. Subfraction vaccines proved to be of low antigenicity and large amounts of expensively produced material would be required to make an effective vaccine. Proponents of such vaccines are working on methods for increasing the antigenicity. We will hear more on this by a member of the Baylor University group later in the conference.

Finally we must say something about the recombinant DNA vaccines. Here I might speculate that such a vaccine may eventually take the place of the current one. Dr. Valenzuela will be speaking on this subject later in the conference. The preparation of a recombinant DNA or synthetic vaccines based on DNA sequence data coding for the surface antigen does not appear to be an insurmountable obstacle any longer. However, these possibilities are still in the future and will require years of laboratory research and clinical trials to be proven safe and less costly than the current vaccine. Regarding the high cost of the present vaccine we must ask the question, "Why the high cost?" In my opinion, high cost can be attributed to one or two features of production. The first is the cost of obtaining carrier blood for the isolation and purification of surface antigen. If cost here is due to scarcity, we must ask why bloods cannot be collected from areas of the world where the carrier rate is known to be high. Payment for units of blood might be welcomed by residents of these areas particularly if it became possible to immunize the population with vaccine at a lower cost. If the major contributing cost is safety testing, and we know that this is costly because it involves chimpanzee testing and a holding period of up to six months, then perhaps we should extend our efforts to finding alternative safety tests. This is not an impossible idea if one considers that there are now hybridizable probes available for determining the presence of DNA in finished vaccines and also a very sensitive DNA polymerase test to determine the presence of infectious hepatitis virus. Can we determine whether these tests, on a product which is manufactured under the most rigid conditions formulated to date for any vaccine, are sensitive enough?

Hepatitis is a significant problem in the medically underdeveloped parts of the world, but even more so in sub-Saharan Africa, parts of Asia and Oceania and elsewhere where infection rates may be extraordinarily high. In these countries chronic liver disease and primary cancer of the liver can be very common and there is reason to believe that much of this is associated with persistent infection with HBV. We expect that the present vaccine will prove to be the first effective vaccine against a deadly cancer in man.

ACKNOWLEDGMENTS

This work was supported by USPHS grants CA-06551, RR-05539 and CA-06927 from the National Institutes of Health and by an appropriation from the Commonwealth of Pennsylvania.

REFERENCES

1. Blumberg, B. S., Gerstley, B. J., Hungerford, D. A., London, W. T., and Sutnick, A. I. (1967). Ann. Intern. Med. 66:924.

2. Bayer, M. E., Blumberg, B. S., and Werner, B. (1968). Nature 218:1057.
3. Coyne, V. E., Blumberg, B. S., and Millman, I. (1971). Proc. Soc. Exp. Biol. Med. 138:1051.
4. Okochi, K. and Murakami. S. (1968). Vox Sang. 15:374.
5. Okochi, K., Murakami, S., Ninomiya, M., and Kaneko, M. (1970). Vox Sang. 18:289.
6. Blumberg, B. S., Sutnick, A. I., London, W. T., and Millman, I. (1971). Critical Reviews in Clinical Laboratory Sciences, Vol. 2, issue 3. Australia Antigen and Hepatitis. A Comprehensive Review, p. 32.
7. Lichter, A. E. (1969). Nature 224:810.
8. Hirschman, R. J., Shulman, N. R., Barker, L. F., and Smith, K. O. (1969). JAMA 108:1667.
9. London, W. T., Millman, I., Sutnick, A. I., and Blumberg, B. S. (1970). Clin. Res. 18:536.
10. London, W. T. (1970). Proc. Natl. Acad. Sci. USA 66:235.
11. London, W. T. and Blumberg, B. S. (1973). Symposium IVth Int. Cong. Primat. 4:30.
12. Krugman, S., Giles, J. P., and Hammond, J. (1967). JAMA 200:365.
13. Krugman, S., Giles, J. P., and Hammond, J. (1970). J. Infec. Dis. 122:432.
14. Soulier, J. P., Blatix, C., and Courouce, A. M. (1972). Am. J. Dis. Child. 123:429.
15. Purcell, R. H. and Gerin, J. L. (1975). Am. J. Med. Sci. 270:395.
16. Hilleman, M. R., Buynak, E. B., and Roehm, R. R. (1975). In Proceedings of a Symposium on Viral Hepatitis, National Academy of Sciences, Washington, D.C., March 17-19, 1975.
17. Buynak, E. B., Roehm, R. R., Tytell, A. A., Bertland, A. U., Lampson, G. P., and Hilleman, M. R. (1976). Proc. Soc. Exp. Biol. Med. 151:694.
18. Murphy, B. L., Maynard, J. E., and LeBouvier, G. L. (1974). Intervirol. 3:378.
19. Barker, L. F., Gerety, R. J., Hoofnagle, J. H., and Nortman, D. F. (1974). In Transmissible Disease and Blood Transfusion, 6th Am. Red Cross Scientific Symp., Washington, D. C., T. J. Greenwalt and D. Q. Jamieson, eds.
20. Maupas, P., Coursaget, P., Goudeau, A., and Drucker, J. (1976). Lancet 1:1367.

CLINICAL EXPERIENCE WITH HEPATITIS B VACCINE

Arlene A. McLean, Eugene B. Buynak, Barbara J. Kuter, Maurice R. Hilleman and David J. West

Virus and Cell Biology Research Department
Merck Sharp and Dohme Research Laboratories
West Point, Pennsylania 19486

INTRODUCTION

The hepatitis B vaccine which is commercially available in the United States consists of noninfectious hepatitis B surface antigen (HBsAg) which is purified from the plasma of asymptomatic chronic carriers of the hepatitis B virus (1,2,3). The excess HBsAg subunit present in large quantities in the plasma of these individuals first is separated from the heavier hepatitis B virus and from extraneous blood proteins by two ultracentrifugation steps. It then is subjected sequentially to three chemical treatments (pepsin at pH 2, 8M urea and formaldehyde) to insure killing of any residual hepatitis B virus as well as any other infectious agent which might be present in human plasma. The vaccine manufacturing facility provides total physical separation of each of the critical inactivation steps and prevents introduction of untreated materials into later process steps. All chemical reagents are assayed for potency before and after use to insure proper inactivation at each of the steps. All lots of vaccine are extensively tested for purity, potency and safety, including a test for safety in susceptible chimpanzees.

Early clinical studies showed the aqueous vaccine to be poorly immunogenic for man (4). Therefore, the purified HBsAg has been adsorbed onto aluminum hydroxide as an adjuvant. A one milliliter dose of the vaccine contains 20 μg of HBsAg, 0.5 mg of aluminum ion and 50 μg of thimerosal preservative. One milliliter is the recommended vaccine dose for healthy adults, and the recommended regimen is two doses a month apart followed by a third or booster dose six months after the first. Children require smaller doses

(10 μg), and persons who are immunocompromised/suppressed requ e larger doses (40 μg). All injections, except for those given t persons with hemophilia or similar bleeding disorders, are admi -stered intramuscularly. This paper reviews the clinical data on which these recommendations were based.

POPULATIONS STUDIED

Between November 1975 and market introduction in June 1982, th hepatitis B vaccine was administered to more than 19,000 volunteers as part of clinical studies conducted in the United States and in 28 countries throughout the world. Of the 19,000 volunteers, more than 11,000 were known to be susceptible to hepatitis B prior to vaccination; 225 had hepatitis B antibodies from previous infection, and 17 were chronically infected with the hepatitis B virus. The remaining 8000 volunteers, mostly medical and laboratory personnel, were vaccinated as part of a demonstration project conducted in Switzerland (5). These individuals were not routinely tested for hepatitis B markers prior to vaccination.

Populations enrolled in the studies are summarized in Table 1. These included healthy infants, children and adults at low risk for infection with hepatitis B virus, as well as persons whose occupations, lifestyles or underlying conditions placed them at increased risk.

Table 1. Persons included in clinical studies of hepatitis B vaccine conducted between November 1975 and June 1982.

Population Group		Number Vaccinated
Medical and laboratory personnel		10,315
Low-risk adults		3731
Homosexually active males		1926
Hemodialysis patients		1572
Healthy infants and children		1328
Family contacts of HBsAg positives		163
Children with thalassemia		100
Mentally retarded		90
Renal transplant patients		78
Hemophiliacs		72
Persons with malignancies		54
Infants born to HBsAg carriers		53
Drug addicts		45
	Total	19,527

VACCINE SAFETY

Essentially all of the more than 19,000 persons who participated in the clinical studies of the vaccine have been carefully monitored for occurrence of adverse reactions to the vaccine. Some of these individuals have been followed for up to 8 years for appearance of late-occurring reactions. All of the data indicate that the vaccine has an excellent safety profile. In fact, no serious adverse reaction has been reported by any volunteer who participated in the clinical studies. There has been no instance in which vaccination caused hepatitis B, and there have been no reports of anaphylactoid or serious neurological reactions to the vaccine.

The types and incidence of complaints which generally are reported within 5 days following vaccination are summarized in Table 2. These data were collected following the administration of 3516 doses of vaccine to 1255 healthy adults who participated in studies conducted principally at the Merck Sharp and Dohme installations at West Point, PA and Rahway, NJ. The most commonly reported complaint was mild soreness at the injection site which was usually limited to the first 48 hours after vaccination. Other injection-site reactions including erythema, swelling, warmth and induration were occasionally reported. Systemic complaints such as headache, fever, upper respiratory symptoms, gastrointestinal illness including nausea, vomiting, abdominal pain and diarrhea, and fatigue each were reported following 2-3% of the injections, while myalgia, malaise and arthralgia occurred after 1% or less of the vaccinations.

Table 2. Frequency of clinical complaints within 5 days following administration of 3516 doses of vaccine to 1255 healthy adults

Clinical complaint	% of doses with complaint
Injection-site reaction	12.3
Headache	3.1
Upper respiratory illness	2.5
Gastrointestinal illness	2.0
Fatigue	1.9
Fever ($\geq$ 100°F, Oral)	1.8
Myalgia	1.2
Malaise	0.8
Arthralgia	0.7

There was no tendency towards increased frequency or severity of complaints with successive doses of the vaccine. In fact, the opposite was true; more of the individuals in these studies reported reactions to the first dose of vaccine than to the second or third doses. Individuals who had an adverse reaction to the first or second dose did not experience more severe reactions when additional vaccine doses were administered.

It is also important to note that findings from four large, double-blind, placebo-controlled studies (6,7,8,9,10) conducted among adults at high risk of contracting hepatitis B showed no significant difference in the overall incidence of adverse reactions occurring in the vaccine group as compared to recipients of alum placebo.

ANTIBODY RESPONSE IN THE HEALTHY INDIVIDUAL

Since the vaccine is a pure preparation of HBsAg, the only antibody which is formed in response to vaccination is anti-HBs; anti-HBc and anti-HBe are not formed. Clinical studies cited below have demonstrated that anti-HBs alone protects against hepatitis B.

In the healthy individual, age is the most important factor influencing the antibody response to the vaccine. The younger the vaccinee, the more likely he is to develop an immune response. Table 3 illustrates this point by showing antibody seroconversion rates by age for persons who received 10, 20 or 40 μg doses of vaccine. In all cases, vaccine was given intramuscularly at time 0, 1 and 6 months, and blood specimens collected 3, 6 and 7 months after the first injection were assayed in the same laboratory (MSDRL) for anti-HBs using a radioimmune assay (AUSAB; Abbott Laboratories). Antibody titers were reported in estimated AUSAB RIA units, and sera with titers less than 8 were considered to be negative. There is a progressive decrease in responsiveness with increasing age for all three of the dose levels tested. In these studies, doses of as little as 10 μg were sufficient to seroconvert 100% (70/70) of individuals in the 1-19 age group. Doses of 20 μg or 40 μg gave almost identical response curves in all age groups, while doses of 10 μg elicited a somewhat slower response in those individuals over 20 years of age. A similar age-antibody response relationship has been noted by Deinhardt (11).

The response in adults is further illustrated in Figure 1 which shows the anti-HBs seroconversion rate by time and antibody titer level (estimated AUSAB RIA units) for 482 low-risk adults, 20 to 70 years old, who received three 20 μg doses of vaccine. These data show that, in adults, antibody develops slowly. By 1 month after the first infection, only 31% of the vaccinees had detectable anti-HBs. By 3 months, the seroconversion rate rose to

Table 3. Antibody (anti-HBs) seroconversion rates according to age and dose of hepatitis B vaccine among healthy individuals.

		Seroconversion rate (%)		
Age	Time	10 μg	20 μg	40 μg
1-19 yrs.	3 mos.	118/121 (98)	12/12 (100)	48/48 (100)
	6 mos.	89/89 (100)	12/12 (100)	47/47 (100)
	7 mos.	70/70 (100)	9/9 (100)	25/25 (100)
20-39 yrs.	3 mos.	151/184 (82)	713/792 (90)	70/75 (93)
	6 mos.	149/169 (88)	707/752 (94)	69/73 (95)
	7 mos.	147/159 (92)	673/695 (97)	71/73 (97)
40-59 yrs.	3 mos.	98/189 (52)	190/264 (73)	38/47 (81)
	6 mos.	121/184 (66)	209/253 (83)	41/48 (85)
	7 mos.	133/182 (73)	221/244 (91)	44/48 (92)
60-79 yrs.	3 mos.	16/35 (46)	26/47 (55)	3/7 (43)
	6 mos.	24/38 (63)	32/47 (68)	4/8 (50)
	7 mos.	27/37 (73)	36/44 (82)	6/8 (75)

81%; and by 6 months, 88% of the vaccinees seroconverted, though only 37% of them had high levels of anti-HBs, i.e. responses of 382 AUSAB units or more (comparable to approximately 100 milli-International Units (mIU) per ml). After the booster dose, the seroconversion rate increased to 92% and nearly 74% of the vaccinees developed high levels of circulating antibody.

Clinical studies have demonstrated that all three vaccine doses are necessary to induce an enduring immune response in adult recipients (12). The first two doses are priming doses which serve to seroconvert the majority of individuals. The third dose seroconverts a small number of vaccinees and, in addition, greatly increases (about 10-fold) the titer levels of those with antibodies. This increase is necessary for long-term persistence of immunity. The timing of the three vaccine doses is also important. The second dose should be given 1 to 2 months after the first to maximize the primary response. The third dose should not be given until approximately 6 months after the first dose when the primary response is complete. It may, however, be delayed for an additional 6 months (12 months after the first injection) without significantly affecting the booster response (C. Stevens, personal communication).

As noted, approximately 5-10% of healthy adults who receive the standard three-dose regimen of 20 μg vaccine fail to develop

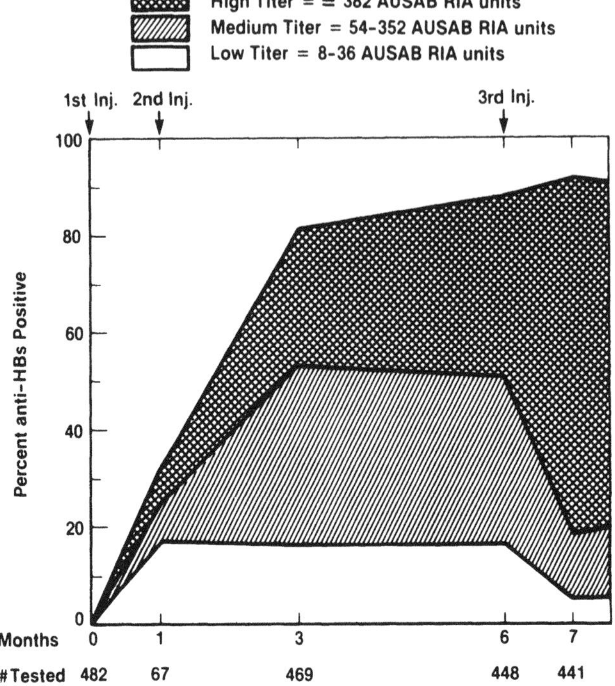

Figure 1. Anti-HBs seroconversion rates and distribution of titers in AUSAB RIA units for healthy adults receiving three 20 μg doses of vaccine. AUSAB RIA units are reported as discrete rather than continuous values as suggested in Abbott's operating procedure for this assay.

anti-HBs. The majority of nonresponders tend to be persons over the age of 40. In a small study of nonresponders, we revaccinated 18 individuals with an additional one to three doses of 20 μg vaccine. Overall, 33% of these nonresponders developed antibody after receiving a total of four to six doses of vaccine (unpublished data).

Long-term persistence of vaccine-induced antibodies is being studied among several populations including low-risk children and adults (12), health care workers (13,14) and homosexually active males. Antibody has been found to persist in 100% of children ages 1-10 four years after receiving three 10 μg doses of vaccine (Weibel, unpublished data). The antibody titers of these children are excellent, persisting at levels nearly 100 times that of healthy adults.

Figure 2 presents findings for a group of healthy adults at low risk of contracting hepatitis B. Titers decreased substantially

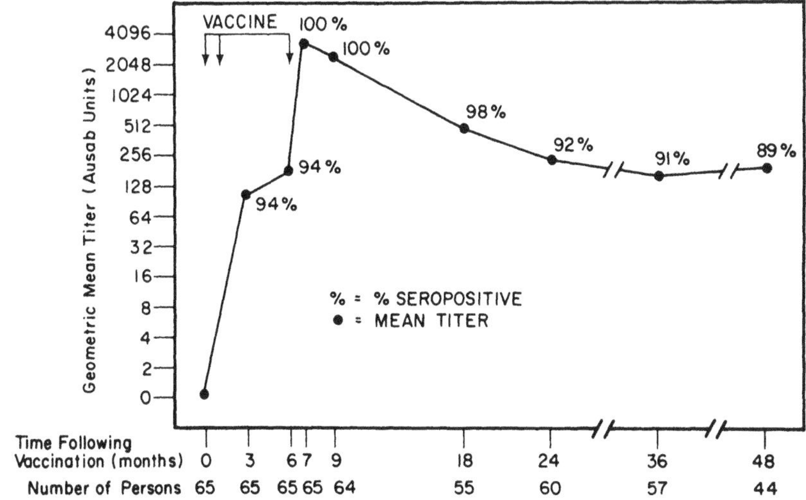

Figure 2. Antibody development and persistence in adults who received three doses, 40 μg of alum hepatitis B vaccine.

during the interval from 7 to 24 months following vaccination but remained fairly stable between 2 and 4 years after vaccination. Importantly, 89% of those who responded to three doses of the vaccine retained detectable anti-HBs 4 years following vaccination.

Almost identical results were obtained for a group of homosexually active males followed for 4 years after vaccination (C. Stevens, personal communication). Most importantly, among those men who lost detectable anti-HBs and then were exposed to hepatitis B, none developed detectable antigenemia or symptoms of hepatitis B. Instead, upon exposure to the virus, all of the men showed a rapid increase in anti-HBs. In some cases anti-HBc also developed, indicating that viral infection had occurred but, in all cases, this infection was subclinical.

PASSIVE-ACTIVE IMMUNIZATION

Studies by Deinhardt (15), Szmuness (16) and Weibel (unpublished) have shown that doses of up to 5 ml of hepatitis B immune globulin (HBIG) do not inhibit the active anti-HBs response in adults who receive simultaneous vaccination with hepatitis B vaccine. The studies showed that anti-HBs was detectable in the serum of almost all individuals within hours following administration of the HBIG and vaccine, and remained detectable for the entire 8-24 months of follow-up. The main advantage of passive-active treatment is that the passively transferred antibodies are effective during the 1- to 3-month interval that is necessary for the individual to mount an active immune response to the vaccine. Such immediate protection may be important in cases of accidental inoculation with contaminated needles or for sexual contacts of HBsAg carriers.

Excellent active antibody responses also are achieved among infants who have high levels of maternally transferred anti-HBs at the time of vaccination (17).

USE IN HIGH-RISK POPULATIONS

The immunogenicity of the vaccine has been tested in many high-risk populations as shown in Table 1. The majority of these populations, including health care professionals (6,14,18,19), homosexually active males (7,8), thalassemics (20), hemophiliacs (21), infants born to carrier mothers, and drug addicts (22) (whether taking methadone or not) respond well to the vaccine. The study in hemophiliacs by Desmyter et al. (21) also indicates that the vaccine is well tolerated when administered to these individuals via the subcutaneous route.

Findings among the mentally retarded have been variable. In a study conducted in Sweden (23) seroconversion to 20 μg doses of vaccine occurred in 12 of 16 (75%) Down's syndrome patients and in 7 of 9 (78%) persons with other types of mental retardation. Findings from a similar study conducted in the Netherlands (Schalm, personal communication) found 90% seroconversion among both the 23 patients with Down's syndrome and the 30 persons with other types of mental retardation, but the antibody titer levels of both groups were lower than that found for a group of healthy controls. A third study being conducted in the U.S. and Canada by Taylor et al. (personal communication) shows seroconversion among 12/13 patients with Down's syndrome and 14/14 persons with other types of mental retardation. Additional studies are under way in an effort to identify the reason for these seemingly disparate results. The bulk of information at this time indicates that mentally disabled individuals can be successfully immunized with 20 μg doses of vaccine.

Studies among dialysis patients also have shown variable results; however, the preponderance of information from these studies indicates that dialysis patients do not respond well to the vaccine. The initial study by Stevens et al. (24) indicated that an 89% seroconversion rate could be attained but that 40 μg doses of vaccine would be needed. Subsequent studies by Dr. Stevens and others (15,25,26) have shown that these conclusions were unduly optimistic. Findings from these additional studies show that age, once again, is a major factor determining responsiveness to the vaccine. In a large study conducted among more than 600 renal dialysis patients who received 40 μg doses of vaccine, the overall seroconversion rate was 70%; but, for those less than 40 years old, the antibody response rate was 85% (Stevens, personal communication).

Only 20-40% of persons who are receiving immunosuppressive therapy, such as renal transplant patients and cancer patients, respond to the vaccine (25,27).

PROTECTIVE EFFICACY

Four large double-blind, placebo-controlled protective efficacy studies (6,7,8,9,10) have been conducted among healthy adults at high risk of infection with hepatitis B virus. These studies have shown that vaccine-induced antibodies protect the vaccinee against acute hepatitis B, asymptomatic infection and chronic antigenemia. Protection is excellent whether the viral challenge is percutaneous or via mucosal surfaces. In one of these studies (9), a monovalent type _ad_ vaccine was shown to afford 92% protective efficacy against disease caused by type _ay_ challenge. Two of these studies (7,10) provided suggestive evidence that vaccination might provide partial protection when begun post-exposure.

Studies to measure vaccine efficacy when administered post-exposure are being conducted among infants born to HBeAg positive mothers. Three studies are under way, and preliminary findings are available from two of these. Beasley has found that 20 μg vaccine alone is 75% efficacious in preventing chronic antigenemia when administered at 1 week and 1 and 6 months after birth (28). Similar results were obtained with 0.5 ml doses of HBIG administered at 0, 3 and 6 months of age (29). Use of both HBIG and vaccine has been shown to be even more efficacious. Passive immunization with a single 0.5 ml dose of HBIG at birth followed by vaccine at 1 week, 1 month and 6 months has been found to be 93% efficacious in preventing chronic hepatitis B infection in high-risk infants (30). Preliminary results from a similar trial by Stevens indicate that a 0.5 ml dose of HBIG at birth plus vaccine at either 0, 1 and 6 months or 1, 2 and 6 months of age is approximately 86% efficacious. These data strongly suggest that perinatal transmission of hepatitis B infection can be prevented through early intervention with both passive and active immunization.

ACKNOWLEDGMENT

Some of this material was presented at the Symposium by Dr. Cladd Stevens.

REFERENCES

1. Buynak, E. B., et al. (1976). Development and chimpanzee testing of a vaccine against human hepatitis B. Proc. Soc. Exp. Biol. Med. 151:694-700.
2. Buynak, E. B., et al. (1976). Vaccine against human hepatitis B. JAMA 235:2832-2834.
3. Hilleman, M. R., et al. (1983). The preparation and safety of hepatitis B vaccine. J. Infect. 7:3-8.
4. Hilleman, M. R., et al. (1978). Clinical and laboratory studies of HBsAg vaccine, in Viral Hepatitis, G. N. Vyas, S. N. Cohen and R. Schmid, eds., Franklin Institute Press, Philadelphia, pp. 525-537.
5. Grob, P. J., et al. (1983). Hepatitis B vaccination of high-risk individuals of the Canton Zurich. Eur. J. Clin. Microbiol. 2:309-315.
6. Dienstag, J. L., et al. (1982). Hepatitis B vaccine in health care personnel: A randomized, double-blind, placebo-controlled trial. Hepatology 2:696.
7. Francis, D. P., et al. (1982). The prevention of hepatitis B with vaccine. Report of the Centers for Disease Control multi-center trial among homosexual men. Ann. Intern. Med. 97:362-366.
8. Szmuness, W., et al. (1980). Hepatitis B vaccine. Demonstration of efficacy in a controlled clinical trial in a high-risk population in the United States. N. Engl. J. Med. 303:833-841.
9. Szmuness, W., et al. (1982). Hepatitis B vaccine in medical staff of hemodialysis units. Efficacy and subtype cross-protection. N. Engl. J. Med. 307:1481-1486.
10. Szmuness, W., et al. (1981). A controlled clinical trial of the efficacy of the hepatitis B vaccine (Heptavax B): A final report. Hepatology 1:377-385.
11. Deinhardt, F. (1982). Control of hepatitis A and B with vaccine. Ann. Clin. Res. 14:236-244.
12. McLean, A. A., et al. (1983). Summary of worldwide clinical experience with H-B-VAX. J. Infect. 7:95-104.
13. Deinhardt, F., et al. (1983). Immune responses to active and passive-active vaccination against hepatitis B. J. Infect. 7:21-25.
14. Larke, R. P. B., et al. (1983). Hepatitis B and the dental profession: Response to hepatitis B vaccine in Canadian dental personnel. J. Infect. 7:27-33.
15. Deinhardt, F. (1983). Aspects of vaccination against hepatitis B; passive-active immunization schedules and immune responses in different age groups. Scand. J. Infect. Dis. (Suppl. 38) 17-23.
16. Szmuness, W., et al. (1981). Passive-active immunization against hepatitis B: immunogenicity studies in adult Americans. Lancet 1:575-577.

17. Prozesky, O. W., et al. (1983). Immune response to hepatitis B vaccine in newborns. J. Infect. 7:53-55.
18. Szmuness, W., et al. (1981). The immune response of healthy adults to a reduced dose of hepatitis B vaccine. J. Med. Virol. 8:123-129.
19. Deinhardt, F., et al. (1981). Passive-active immunization against hepatitis B, in Proceedings of the European Symposium on Hepatitis B, S. Krugman and S. Sherlock, eds., KRP/Infor Media, New York (for MSDI), pp. 140-150.
20. Matsaniotis, N., et al. (1983). Immunogenicity and efficacy of hepatitis B vaccine in normal children and in patients with thalassemia. J. Infect. 7:57-61.
21. Desmyter, J., et al. (1983). Hepatitis B vaccination of hemophiliacs. Scand. J. Infect. Dis. (Suppl. 38): 42-45.
22. Schalm, S. W., et al. (1983). Immune response to hepatitis B vaccine in drug addicts. J. Infect. 7:41-45.
23. Wahl, M., et al. (1983). Immune responses to hepatitis B vaccine in the mentally retarded. J. Infect. 7:47-51.
24. Stevens, C. E., et al. (1980). Hepatitis B vaccine: Immune responses in hemodialysis patients. Lancet 2:1211-1213.
25. Grob, P., et al. (1983). Hepatitis B vaccination of renal transplant and hemodialysis patients. Scand. J. Infect. Dis. (Suppl. 38):28-32.
26. Bergamini, F., et al. (1983). Immune response to hepatitis B vaccine in staff and patients in renal dialysis units. J. Infect. 7:35-40.
27. Arnold, W., et al. (1983). Vaccination of children with malignant diseases with an alum-adsorbed hepatitis B vaccine - immunogenicity studies. Scand. J. Infect. Dis. (Suppl. 38):33-35.
28. Beasley, R. P., et al. Efficacy of HBV vaccine for the interruption of the perinatally transmitted hepatitis B virus carrier state. N. Engl. J. Med. (in press).
29. Beasley, R. P., et al. (1983). Efficacy of hepatitis B immune globulin for prevention of perinatal transmission of the hepatitis B virus carrier state: Final report of a randomized double-blind, placebo-controlled trial. Hepatology 3:135-141.
30. Beasley, R. P., et al. (1983). Prevention of perinatally transmitted hepatitis B virus infections with hepatitis B immune globulin and hepatitis B vaccine. Lancet 2: 1099-1102.

HEPATITIS B VACCINE TRIALS, EXPERIENCE AND REVIEW

Robert E. Weibel

The Joseph Stokes, Jr. Research Institute
The Children's Hospital of Philadelphia
Philadelphia, Pennsylvania 19104

Demonstrating the relationship between Australia antigen and hepatitis B virus in the late 1960's created a serologic means for detecting hepatitis B infection, disease, carriers, and recovery (1,2). This information provided an impetus for vaccine development even in the absence of a means to propagate the virus in the laboratory. Individuals chronically infected with hepatitis B could be identified and their infected plasma containing infectious hepatitis B virus (HBV) or Dane particles and excess hepatitis B surface antigen (HBsAg) obtained by plasmaphoresis (3). Additional markers (shown in Figure 1) of human hepatitis B infection and recovery were further identified as important serologic markers. These include hepatitis B surface antigen (HBsAg) and its antibody (anti-HBs), hepatitis B core antigen and its antibody (anti-HBc), and the hepatitis B e antigen and its antibody (anti-HBe). By use of these and other tests, it is possible to identify susceptibles from immunes and the probable state of infection of the carriers. Persons who have never had hepatitis B are free of all markers.

Epidemiologic studies in the early 1970's identified certain geographic areas of Africa, Asia, South America and Southern Europe as highly endemic for hepatitis B virus, and certain segments of the population including active male homosexuals with many partners at increased risk for life-threatening hepatitis B infection (3). Fulminant hepatitis, chronic active and persistent hepatitis, cirrhosis, immune complex disease and primary hepatocarcinoma have been associated with hepatitis B infection.

Based on the premise that circulating antibody to hepatitis B surface antigens could prevent infection of the liver, efforts

HBV	Hepatitis B Virus - The Dane Particle
HBsAg	Hepatitis B Surface Antigen - Australia Antigen
HBcAg	Hepatitis B Core Antigen
HBeAg	The "e" antigen associated with the core of the Dane particle is present during the most infectious period.
Anti-HBs	Antibody to Hepatitis B Surface Antigen
Anti-HBc	Antibody to Hepatitis B Core Antigen
Anti-HBe	Antibody to the "e" Antigen

Figure 1. Designations of hepatitis antigenic markers and their antibodies.

were begun to obtain large quantities of purified hepatitis B surface antigen free of infectious HBV from infected human plasma. Infected human plasma containing plasma components, 42 n hepatitis B virus or Dane particles and excess hepatitis B surface antigen as 22 n spheroids or aggregated tubules are shown in Figure 2. Groups at Merck, Sharp and Dohme Research Laboratories and the National Institute for Allergy and Infectious Diseases developed laboratory procedures to purify and concentrate hepatitis B surface antigen (4).

Present vaccine preparation (Figure 3) at the Merck Laboratories takes 65 weeks and consists of a sequence of steps that begins with the removal of fibrin from plasma with calcium. Concentration of surface antigen is followed by ammonium sulfate precipitation, isopycnic banding with sodium bromide and sucrose gradient rate zonal sedimentation in an ultracentrifuge. The partially purified antigen concentrate is then digested with pepsin at pH 2 and the antigen is unfolded in 8 molar urea solution to facilitate the removal of extraneous human liver and blood plasma components. After gel filtration the antigen is treated with formalin in 1:4000 dilution, adsorbed onto aluminum hydroxide and preserved with thimerosal. These steps, which include digestion with pepsin, treatment with 8 molar urea and formaldehyde and various physical separation procedures, inactivate all known classes of viruses. Purified surface antigen treated to remove extraneous materials and to destroy infectivity for use in the vaccine is seen in Figure 4.

Prior to the addition of aluminum hydroxide, laboratory and animal tests for safety of the vaccine, shown in Figure 5, are conducted. They consist of *in vitro* and *in vivo* assays of viral

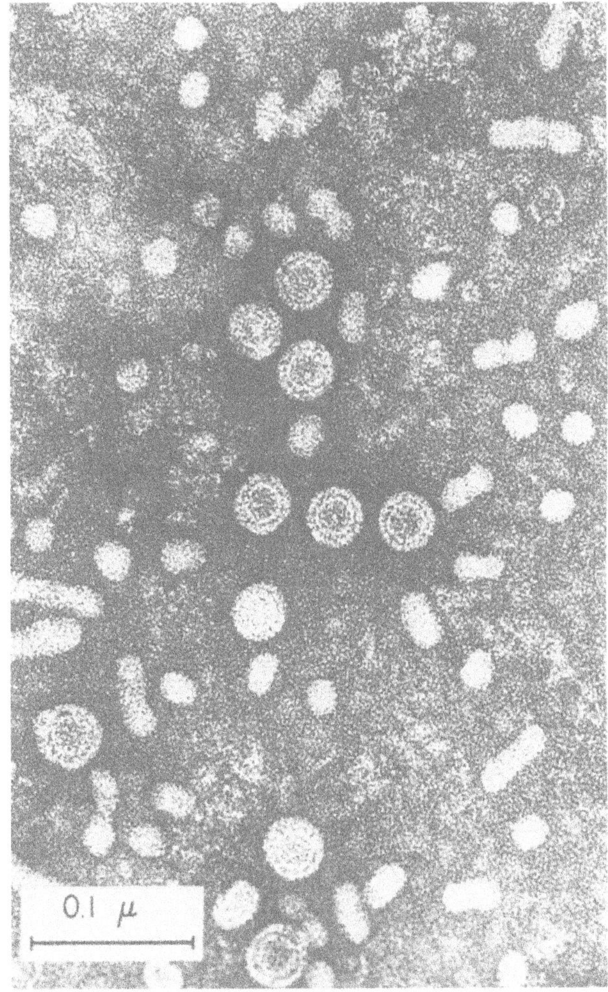

Figure 2. Morphologically diverse 42 n and 22 n particles in hepatitis B positive plasma.

and microbial sterility of bulk plasma, purified antigen and final product states. A six-month safety test is conducted in chimpanzees. All protein in the final vaccine is accounted for as HBsAg measured by quantitative radioimmune assay.

Hepatitis B is a life-threatening disease and safety must be a prime consideration in vaccine development. Before studies in man were initiated with purified hepatitis B vaccine, the vaccine was demonstrated safe and immunogenic in chimpanzees. Hepatitis B

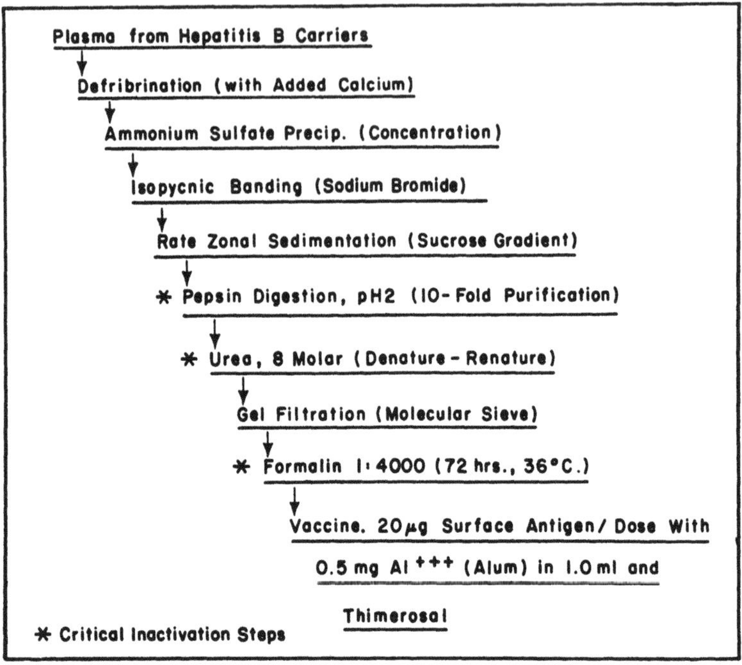

Figure 3. Production of hepatitis B vaccine (Merck, Sharp and Dohme).

virus infects susceptible chimpanzees resulting in mild disease. In addition, the vaccine was shown to protect susceptible chimpanzees when later challenged with hepatitis B infectious serum in a controlled trial. The results of a single animal vaccinee and control are shown in Figure 6. Protective efficacy in the various chimpanzee studies has been 100% (4).

The hepatitis B surface antigen is a complex polypeptide with a common dominant "a" antigen and shared subtypes called d and y and w (1,2,3,4) and r. Studies by several workers have shown that cross protection was afforded when chimpanzees were vaccinated with serotype adw vaccine and challenged with ayw virus and vice versa. Similar cross protection was afforded when immunity in the chimpanzees was established by active infection. Cross protection is considered to be induced by the "a" antigen which is present in all serotypes.

In November 1975, Dr. Saul Krugman administered the first purified aqueous hepatitis B vaccine in the United States to 11 persons. These persons were at low risk for hepatitis B infection and remained free of infection (4). In 1976, in cooperation with Drs. Buynak, McLean and Hilleman, I administered three monthly doses of aqueous vaccine, 20 mcg/dose, to 10 persons. Clinical

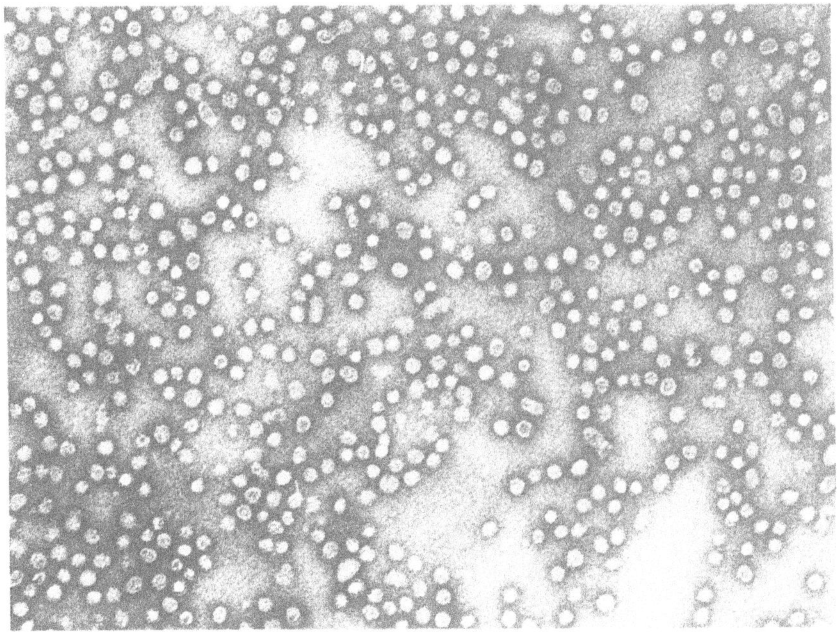

Figure 4. Purified 22 n particles used in the vaccine.

reactions and antibody responses were minimal with no serologic evidence of infection by hepatitis B virus markers or elevated aminotransferases.

In 1977, after many trials, a 20 mcg dose of alum-adsorbed vaccine administered at times 0, 1 and 6 months was considered optimal as seen in Figure 7. Eighty to ninety percent of normal healthy adults developed hepatitis B surface antibody after two doses of vaccine, and more than 95% of persons had antibody after the third dose. The seroconversion rates following 20 or 40 mcg of antigen per dose were similar. Two doses of vaccine in most individuals stimulated substantial antibody responses to afford protective immunity. However, the booster dose given at six months effected an increase in the seroconversion rate and a great increase in the antibody titer level. High antibody titers are followed by long-term immunity (5).

In 1978, Szmuness and Stevens et al. initiated a double blind, controlled efficacy trial among 1083 male homosexuals in New York (6).

In 1980, all 40 children ages 1 to 10 years, administered vaccine demonstrated a far greater antibody response than do

PLASMA POOL

ANIMAL TESTS

Adult mouse (i.p. & i.c.)
Suckling mouse (i.p. & i.c.)

CHICK EMBRYO

Yolk sac
Allantoic sac

CELL CULTURE

Grivet monkey kidney (Vero)
WI-38 (Human diploid)

PURIFIED INACTIVATED BULK ANTIGEN (40 μg/ml)

Microbial sterility
Blood group substance
Human IgM (Immunodiffusion)
Lowry protein - with RIA antigen assay
Formaldehyde
Rabbit pyrogen
Chimpanzee safety
Mouse potency

FINAL CONTAINER TESTS

Microbial sterility
Mouse and guinea pig - General safety
Free formaldehyde
Thimerosal
Alum
Identity
Potency - Quantitative radioimmune assay

Figure 5. Laboratory and animal safety tests for hepatitis B vaccine production.

adults. A 10 mcg dose of vaccine at times 0, 1 and 6 months induced titers 10 times those induced by a 10 mcg dose of vaccine in adults. Further studies with 41 infants 3 to 12 months of age with two 10 mcg doses of vaccine given one month apart demonstrated high antibody titers in all infants. These titers continued to rise at 9 months post-vaccination to a geometric mean titer by radioimmune assay of 10,000.

Responses of all 14 newborns vaccinated within 36 hours of birth with two 10 mcg doses of vaccine one month apart were excellent when assayed 3 to 10 months later.

Persons with hemophilia or thalassemia and drug addicts all respond very well to the vaccine. Preliminary data indicate that the mentally retarded also show good response. Immunocompromised

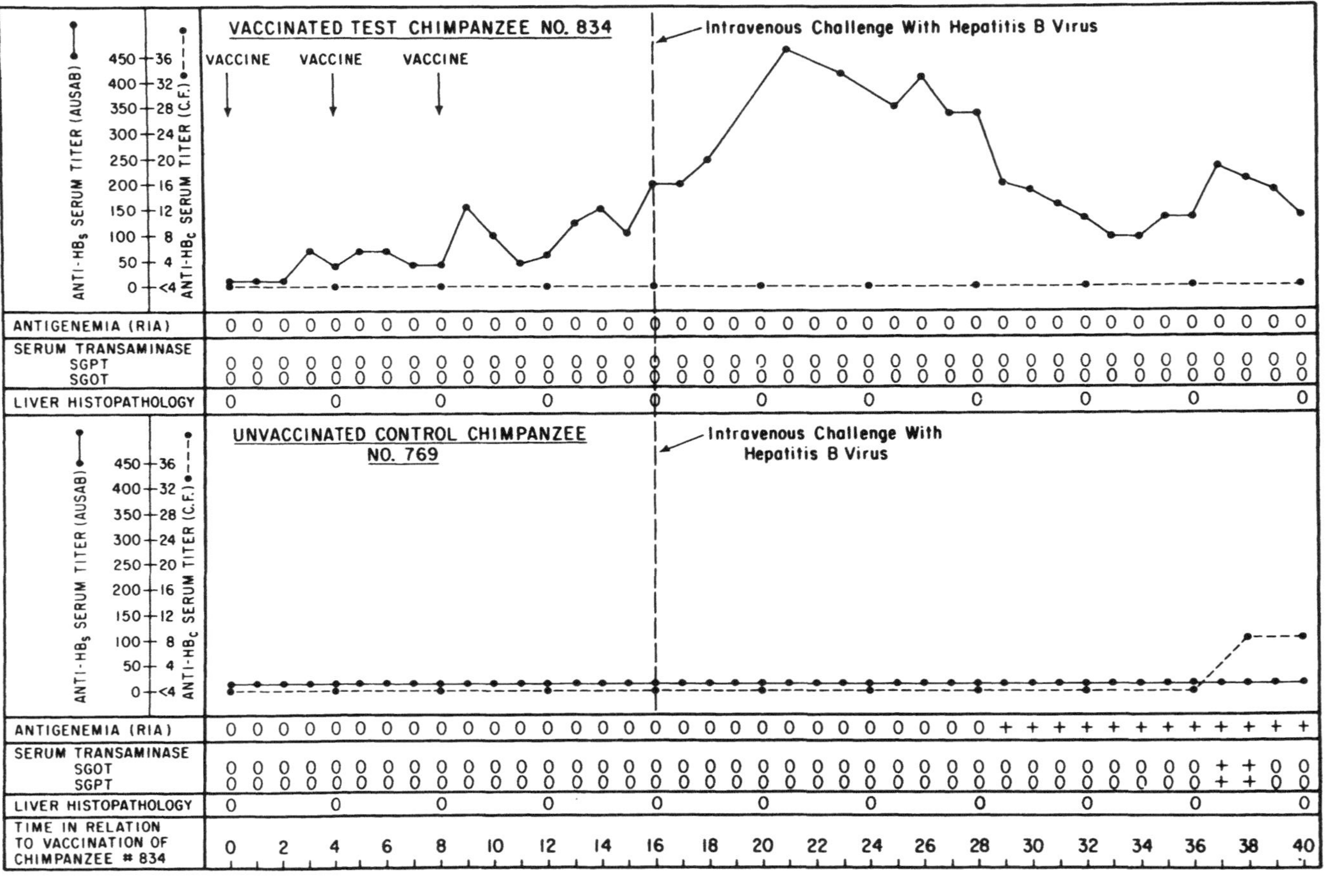

Figure 6. Dynamics of viral, pathologic and serologic responses in a vaccinated and a control chimpanzee before and after i.v. challenge with human hepatitis B virus.

NORMAL ADULTS: 20 μg at time 0, 1 month and 6 months

1 Dose: 30% seroconversion (at 1 month)
2 Doses: 80-90% seroconversion
3 Doses: 95% seroconversion

NORMAL CHILDREN: 10 μg at time 0, 1 month and 6 months (1-10 years old)

95-100% seroconversion to 2 doses

INFANTS: 10 μg at time 0, 1 month and 6 months (Birth - 1 year old)

1 Dose: 57% seroconversion (at 1 month)
2 Doses: 93% seroconversion
3 Doses: In progress

IMMUNOCOMPROMISED AND IMMUNOSUPPRESSED:

(Renal dialysis, renal transplant and cancer patients)
40 μg at time 0, 1 month and 6 months
Response decreases with age and degree of immunosuppression

HEMOPHILIACS AND THALASSEMICS: 10 or 20 μg at time 0, 1 month and 6 months

Respond like normal children

DRUG ADDICTS: 20 μg at time 0, 1 month and 6 months

Respond like normal adults

MENTALLY RETARDED: Under study

Figure 7. Antibody response to vaccination in various groups of seronegative individuals.

and immunosuppressed individuals such as renal dialysis patients, renal transplant patients and leukemia/lymphoma patients do not respond well. Forty microgram doses of vaccine are required and, even then, seroconversion rates hover in the 50 to 80% range.

The younger the individual, the greater the likelihood he will respond, regardless of vaccine dose. The older the person and the greater his immunosuppression, the lower his chances of developing an immune response. The response based on increasing age appears to be a slow decrease from the young child to the elderly. Using 40 years of age as a break point, reponses according to age and dose of vaccine are shown in Figure 8. After two doses of vaccine at 3 months the seroconversion rate for older individuals 40-79

Age	Time	Seroconverted/Total (%)		
		10 µg	20 µg	40 µg
10-39 years	3 mos.	101/114(89%)	614/670(92)	74/79(94)
	6 mos.	89/ 97(92)	569/598(95)	73/77(95)
	7 mos.	63/ 65(97)	489/498(98)	73/75(97)
40-79 years	3 mos.	28/ 38(74)	116/173(67)	41/54(76)
	6 mos.	18/ 23(78)	120/153(78)	45/56(80)
	7 mos.	16/ 18(89)	120/134(90)	50/56(89)

Figure 8. Seroconversion of different age groups as a function of dose of hepatitis B vaccine.

years is 15% less than the younger group 10-39 years at all time periods assayed. Aging of the immune system is associated with a less vigorous antibody response.

CLINICAL REACTIONS

Clinical reactions associated with the vaccine are minimal. In the controlled clinical trials by Dr. Szmuness and associates, soreness at the injection site was reported 5% more frequently and low grade fever 0.5% more frequently among vaccinees than placebo recipients (6). Dr. Francis and associates comparing vaccinees and placebo recipients observed about a 2% increase in injection site soreness of less than 1 day among vaccinees (7).

In our experience, reacion associated with alum-adsorbed hepatitis B vaccine was minimal. Following administration of hepatitis B vaccine to 395 adults, a 6% excess of low grade fever (99-100° oral) was reported at 4 hours compared to day 5. Local injection site complaints, usually soreness, were reported by about 20% of adults. Children receiving the 10 mcg dose of vaccine had no consistent temperature elevations and 10% reported injection site soreness, erythema or ecchymosis. Complaints among all groups were more frequent following the first dose of vaccine and least following the third dose. No serious vaccine related reaction or hepatitis B infection was observed.

As of March 1982 (Figure 9), Merck had prepared 19 separate lots of vaccine. All 19 lots had been tested in chimpanzees, and 13 lots had been tested in man without adverse effect. More than 6000 initially seronegative and 225 seropositive persons had received the vaccine given intramuscularly, all without untoward effect. To date, since 1975, over 19,000 adults, children and newborns have received hepatitis B vaccine intramuscularly in clinical trials. Hepatitis B antigenemic carriers were neither

19	lots tested in chimpanzees without untoward effect
13	lots tested in man without untoward effect
6000	seronegative persons vaccinated without untoward effect
225	seropositive persons vaccinated without untoward effect
17	HBsAg carriers vaccinated without untoward effect
7500	Persons vaccinated as part of a demonstration project

Figure 9. Safety considerations for hepatitis B vaccine.

helped nor harmed by vaccine. Reactions to the vaccine itself are principally mild and self-limited and consist of local inflammation at the injection site (15-20%), rarely with fever (3%). There have been no long-term complaints or infections causally related to the vaccine.

PROTECTIVE EFFICACY

The first efficacy trial carried out by Dr. Szmuness, Stevens and colleagues in New York among homosexual men demonstrated vaccine protection determined by analysis of life table attack rates following each dose of vaccine. Analysis in terms of clinically apparent hepatitis B disease or both apparent and inapparent hepatitis infections in the vaccinated and placebo groups during the 735 days of observation following vaccination is shown in Figure 10. It is seen that protection against disease and against inapparent infections exceeded 90% after the first dose of vaccine. Protection against disease was 100% after the third dose of vaccine, 95% against all infections (6).

There was evidence also that the vaccine was partially effective when given post-exposure either in preventing disease or in reducing the severity of the disease in some but not all persons.

A second trial of human hepatitis B vaccine was carried out by Dr. Francis and his colleagues at the Centers for Disease Control. The blind, placebo-controlled study, involved 1400 male homosexuals who received either placebo or three 20 mcg doses of vaccine. Their study showed high level protective efficacy and that the induction of anti-HBs antibody was synonymous with protection against the disease. Protective efficacy in the order of 97% was shown among the antibody responders (7).

Endpoint	Placebo Group No. events	Placebo Group LTAR*	Vaccine Group No. events	Vaccine Group LTAR*	Protective Efficacy (%)
1. Hepatitis B:					
following 1st dose	57	17.0	6	1.2	91.8
following 2nd dose	53	16.3	5	1.0	93.9
following 3rd dose	49	15.6	0	0	100.0
2. All HBV infections, excl. anti-HBc conversions:					
following 1st dose	88	24.6	11	2.4	90.2
following 2nd dose	82	23.7	10	2.2	90.7
following 3rd dose	69	21.5	4	1.1	94.9

* Life-table attack-rate

Figure 10. Protective efficacy rates as a function of number of vaccine doses (ref. 8).

Additional efficacy studies among renal dialysis patients whose immune responses are impaired by disease or immunosuppressive therapy will be needed to establish an effective regimen. Another study conducted by Dr. Jules Dienstag among 1300 physicians and nurses in the Boston area was discontinued for ethical reasons after the vaccine was licensed before statistical efficacy could be demonstrated. Vaccine prepared in France has been demonstrated effective in preventing hepatitis in children.

One problem in handling known exposures to human hepatitis B infection is to provide both immediate and long-term protection by combining passive with active immunization. This method is already employed in tetanus and rabies control programs and has been studied by Deinhardt and associates in Germany and by Szmuness and others in the United States. Their results among adults indicate that passively acquired antibody did not interfere with an active immune response to the vaccine administered simultaneously or one month later. This kind of regimen is also being studied in newborn infants who are born to mothers who are carriers of hepatitis B infection. This combination treatment might be used in cases of accidental inoculations with contaminated medical instruments and perinatal or sexual exposure to hepatitis B virus (9).

DURATION OF ANTIBODY

Persistence of antibody and immune memory is an important measure of protection against hepatitis B infection. In Figure 11, antibody has been shown to persist for at least three years in most persons, achieving a plateau at a lower mean antibody level. The decline in mean antibody titer would suggest the need for a single dose revaccination at five to ten years after the initial immunization.

The response to a single 20 mcg dose revaccination in previously vaccinated persons who had lost their antibodies has been measured to determine their response. Twelve of the 13 persons studied responded anamnestically within one month. One person failed to respond to revaccination. This immediate boost of antibodies following a single dose of vaccine indicated persistent immunologic memory which may be retained long after vaccine induced antibody can be measured (Figure 12).

RECOMMENDATION FOR IMMUNIZATION

Based on the safety and efficacy of hepatitis B vaccine in clinical trials, the Advisory Committee on Immunization Practices of the Centers for Disease Control has recommended immunization of persons at high risk for acquiring hepatitis B infection (9) (Figure 13). High risk groups would include certain health care workers, dialysis patients, institutionalized mentally retarded children and staff, homosexually active males, users of illicit drugs and young refugees from countries with high rates of hepatitis B. In general, persons who handle blood and blood products or are at risk of accidental or inadvertent exposure to hepatitis B virus are included on the list to receive hepatitis B vaccine.

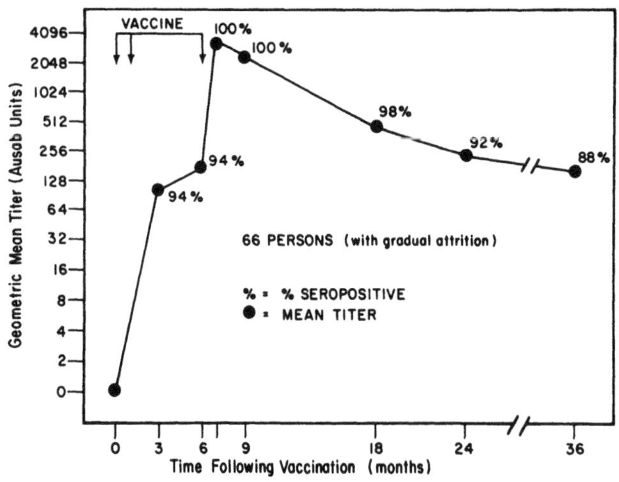

Figure 11. Antibody development and persistence in adults who received three doses, 40 ug each, of alum precipitated hepatitis B vaccine.

Case No.	Vaccination		Revaccination	
	Pre	Max. Titer Post	Pre	1 Mo. Post
46	<8*	412	<8	54
100	<8	477	<8	3600
116	<8	160	<8	72
141	<8	292	<8	540
156	<8	2350	<8	720
168	<8	360	<8	720
205	<8	360	<8	1580
269	<8	16	<8	382
273	<8	54	<8	<8
283	<8	160	<8	720
302	<8	16	<8	3600
316	<8	80	<8	540
320	<8	160	<8	1580

* Estimated AUSAB Units.

Figure 12. Anamnestic response in previously vaccinated persons who received a single 20 μg booster dose.

How best to utilize hepatitis B vaccine in a cost effective manner is now the responsibility of the medical community. The proper use of this highly effective vaccine for pre-exposure protection should markedly reduce and eventually eliminate hepatitis B virus transmission.

Health Care workers (Medical, Dental, Laboratory)

Selected patients in hemodialysis and hematology/oncology units

Children with thalassemia and hemophilia

Residents and staff of institutions for the mentally handicapped

Household contacts of carriers

Classroom contacts of carriers

Homosexually active males

Female prostitutes

Users of illicit drugs

Prisoners

Certain military personnel

Infants and young children immigrating from high endemic areas

Figure 13. High risk groups for hepatitis B infection in the United States.

REFERENCES

1. Prince, A. M. (1968). An antigen detected in the blood during the incubation period of serum hepatitis. Proc. Natl. Acad. Sci. 60:814-821.
2. Krugman, S., et al. (1971). Viral hepatitis, type B (MS-2 strain). Studies on active immunization. J. Am. Med. Assoc. 217:41-45.
3. Vyas, G. N., Cohen, S. N., and Schmid, R. (1978). Viral Hepatitis: A Contemporary Assessment of Etiology, Epidemiology, Pathogenesis and Prevention, Franklin Institute Press, Philadelphia.
4. McAuliffe, V. J., et al. (1980). Type B hepatitis: A review of current prospects for a safe and effective vaccine. Rev. Infect. Dis. 2:470-492.
5. Hilleman, M. R., McAlleer, W. J., and McLean, A. A. (1981). Human Hepatitis B Vaccine, Proc. Europ. Symp. on Hepatitis B, KPR Inform/Media, New York, pp. 120-139.
6. Szmuness, W., et al. (1978). Hepatitis B vaccine. Demonstration of efficacy in a controlled clinical trial in a high risk population in the United States. N. Engl. J. Med. 303:833-841.
7. Francis, D. P., et al. (1982). The prevention of hepatitis B with vaccine. Report of the Centers for Disease Control multi-center efficacy trial among homosexual men. Ann. Intern. Med. 97:362-366.
8. Szmuness, W., Alter, J. H., and Maynard, J. E. (1981). Viral Hepatitis, 1981 International Symposium, Franklin Institute Press, Philadelphia.
9. Morbidity and Mortality Weekly Report 31:317-322; 327-328, 1982.

PRIORITIES FOR THE USE OF HEPATITIS B VACCINE

Roger H. Bernier[1], Mark A. Kane[2], Neal Nathanson[3] and Donald P. Francis[2]

[1]Epidemiologic Research Section, Surveillance, Investigations and Research Branch, Division of Immunization, CDC, Atlanta, GA

[2]Hepatitis Branch, Center for Infectious Diseases, CDC, Atlanta, GA

[3]Department of Microbiology, University of Pennsylvania School of Medicine, Philadelphia, PA

INTRODUCTION

General recommendations for the use of hepatitis B virus (HBV) vaccine in the United States have been formulated by the Public Health Service's Immunization Practices Advisory Committee (the ACIP) (1). The recommendations are based on information about the safety and efficacy of the vaccine and on epidemiologic data. The purpose of our presentation is to review the epidemiologic background of hepatitis, describe the existing ACIP recommendations and the issues they have raised, and assess the likely impact of the current approach in achieving control of HBV-related disease.

BACKGROUND

The prevalence of HBV markers among the general populations of various countries ranges from a low of 5-10% in some countries of Europe and North America to a high of almost 100% in certain developing countries of Africa and Asia (2). The United States as a whole is considered a low prevalence area, but the burden imposed by hepatitis B is considerable. According to data collected by CDC's Hepatitis Branch approximately 20,000 cases of viral hepatitis B were reported in 1980. Individuals between the ages of 15-29 accounted for the largest number (3). CDC estimates that approximately ten times that number, or 200,000, hepatitis B infections occur annually in the United States. One quarter of these persons become ill with jaundice, more than 10,000 patients are hospitalized, and an average of 250 die of acute fulminant disease (1). Between 6-10% of young adults who develop HBV infections become carriers, and an estimated 12,000-20,000 new carriers are added each year to an existing pool estimated to number between 400,000-800,000. Chronic active hepatitis is expected to develop in approximately one quarter of these individuals. Furthermore, an estimated 4000 persons die from HBV-related cirrhosis each year, and more than 800 die from HBV-related liver cancer (1).

HBV is principally a blood-borne virus but is also present in other body fluids such as semen, saliva and serous secretions. Several routes of transmission have been documented, including percutaneous routes associated with needle use in medical and nonmedical settings, and by means of less apparent percutaneous lesions and through mucous membranes. These routes are important in situations of close contact such as occurs in households and in certain types of institutions. Transmission associated with sexual contact now accounts for more infections than needle use (4). Both acute and carrier infections are potentially infectious, but the carrier state serves as a reservoir of HBV and is of paramount importance to the perpetuation of HBV in the population.

VACCINE STRATEGY

The development of a safe and effective HBV vaccine as described earlier during this symposium has been rightly hailed as a major scientific achievement. Undoubtedly, many individuals will benefit. The effective control of HBV transmission on a population basis, however, remains to be accomplished. Equivalent success in the public health arena will depend on the development of an effective strategy for use of the vaccine and on the adequate implementation of this strategy. Ideally, what are the options? A fundamental choice is that between universal versus selected vaccination. In high prevalence areas HBV infection is so widespread that universal vaccination is the only option that can be contemplated as a means of controlling HBV-related diseases. Unfortunately, the high cost and limited quantity of the first generation hepatitis B vaccines make the adoption of such a strategy impossible at the present time.

In low prevalence areas such as the United States, the lifetime cumulative risk of HBV infection is 5-10%. Infections are concentrated in subgroups who are at increased risk because of their illness, occupation, life style or ethnic background. Also, not all high-risk groups are at equally high risk. Consequently, the option of a limited vaccination strategy has been adopted in the United States, based on a low overall prevalence and on the existence of identifiable high-risk groups. Priorities for the use of HBV vaccine are established principally on the varying levels of increased risk in the different candidate populations and on cost considerations. The recent ACIP statement on HBV vaccine gives the prevalence of serologic markers of HBV infection in various subgroups of the United States population and ranks them as high, intermediate, or low risk (Table 1).

Table 1

	Prevalence of serologic markers of HBV infection	
	HBsAg(%)	All markers (%)
High Risk		
Immigrants/refugees from areas of high HBV endemicity	13	70-85
Clients in institutions for the mentally retarded	10-20	35-80
Users of illicit parenteral drugs	7	60-80
Homosexually active males	6	35-80
Household contacts of HBV carriers	3-6	30-60
Patients of hemodialysis units	3-10	20-80
Intermediate Risk		
Prisoners (male)	1-8	10-80
Staff of institutions for the mentally retarded	1	10-25
Health-care workers		
Frequent blood contact	1-2	15-30
Low Risk		
Health-care workers		
No or infrequent blood contact	0.3	3-10
Healthy adults (first-time volunteer blood donors)	0.3	3-5

RECOMMENDATIONS FOR THE USE OF HEPATITIS B VACCINE

High-Risk Groups

Immigrants/Refugees

Immigrants and refugees from highly endemic countries have HBV prevalence rates which are among the highest in the world. For example, Indochinese refugees admitted to the United States have an antigen prevalence rate of 13%, and 70-85% of this group have some type of HBV marker (1). There have been approximately 350-400,000 immigrants admitted to the United States annually in recent years, excluding Indochinese refugees (5). Approximately 150,000 of these immigrants have come from Asia and Africa, and assuming a 10% antigen prevalence rate, the number of new carriers added to the United States pool annually from immigration is approximately equal to the number of carriers generated from infections occurring among United States residents. If the 620,000 Indochinese refugees admitted to the United States since 1975 are included in these estimates (6), the total number of new carriers resulting from immigration is almost double that resulting from indigenous infections over the same time period. The ACIP has stated that such high-risk populations deserve special attention, and depending on specific epidemiologic and public health considerations, more extensive vaccination programs may be warranted (1).

What are the issues in achieving control of HBV transmission in this high-risk group? The primary consideration is selection of the target group for vaccination. Since there is widespread agreement that only universal vaccination in childhood could possibly control HBV infection in highly endemic countries, such a strategy would seem equally valid for persons who have migrated to the United States. Conceivably, one could urge a more limited strategy designed to identify carriers and their susceptible contacts. However, recent arrivals to the United States frequently live in crowded conditions and are members of extended households. Universal vaccination seems better justified epidemiologically, but economic constraints may force the selection of priority candidates for vaccination.

A selective approach requires screening programs to identify the pregnant women or the households at greatest risk. A few local health departments have undertaken selective antigen screening programs, particularly among pregnant women as recommended by CDC, but this effort is not widespread. The State of California has recommended routine prenatal screening among women from populations with high antigen prevalence rates, and California's Medi-Cal program will pay for hepatitis B immune globulin (HBIG) and hepatitis vaccine for newborns when indicated (7).

Alaskan Eskimos

A second high-risk general population is the Alaskan Eskimo (1). According to data collected by CDC's Arctic Investigations Program, the prevalence of hepatitis B surface antigen in certain Eskimo villages of southwest Alaska is very high (8). However, the prevalence can be quite variable with nearby villages exhibiting rates of only 1-2%. Community incidence rates ranging from 3-34% per 100 person years have been documented. The highest rate of new infections occurs in villages where antigen prevalence rates are 8-10%. The ACIP, as in the case of refugees and immigrants from highly endemic areas, has stated that this native American population deserves special attention, and that more extensive vaccination programs may be warranted. As with the refugee population, a policy of universal vaccination could be justified in southwest Alaska on the basis of epidemiological evidence. The Indian Health Service has purchased large quantities of vaccine for use in the general population over the next year or two. Priorities for the use of this vaccine will be established by serosurveys to identify antigen prevalence levels. Universal vaccination will be carried out in villages where prevalence rates of antigenemia are high. Screening programs for all pregnant women in this population are already being carried out. Approximately 3% are antigen positive, and newborns of carrier mothers will receive hepatitis B immune globulin and vaccine.

Clients/Staff in Institutions for the Mentally Retarded

Hepatitis B has been recognized as an important health problem in institutions for the retarded for many years. Antigen prevalence has been estimated at between 5-20%, and overall prevalence of any HBV marker has ranged between 35-80% (1). The ACIP recommends HBV vaccine for all susceptible clients and selected staff of institutions for the mentally retarded. This is being interpreted to mean that any staff members in direct contact with clients should be vaccinated. Most of the institutions for the mentally retarded are supported by limited state funds, and adequate funding to vaccinate all of the children and staff may be difficult to obtain. Failure to control HBV transmission in these settings has implications for vaccination policies in society at large. Institutions for the mentally retarded now have mandated requirements to deinstitutionalize as many children as possible. Approximately 5-20% of these children re-entering society are chronic carriers of HBV, and they are enrolled in special education classes in public school systems throughout the country along with other handicapped children. Some of these deinstitutionalized children have special behavioral and medical problems which, if they are carriers of HBV, might allow them to transmit hepatitis B virus. The ACIP has addressed this problem by stating that vaccine may be offered to classroom contacts of deinstitutionalized mentally retarded HBV carriers who behave aggressively or are otherwise likely to expose others to their blood or serous secretions. This is an emotionally charged issue for teachers, staff and parents. Confrontations have led to court battles seeking exclusion of HBV carriers from public school settings. The availability of HBV vaccine can potentially alleviate these difficulties. Screening programs to identify which deinstitutionalized children are carriers are being carried out in many states. In areas where close monitoring of the movement of carriers in large school systems is not feasible, a policy of universal vaccination for all contacts in special education classes might be considered. In smaller systems, vaccination might be restricted to the contacts in classrooms with carriers. Finally, areas with the most limited resources may wish to further characterize carriers according to their HBeAg status and to vaccinate only those classrooms with HBeAg positive carriers who exhibit the pertinent behavioral and medical problems likely to make them transmitters. In New York City, some public funds will be used to purchase vaccine for staff in special education classes.

Users of Illicit Injectable Drugs

Users of illicit injectable drugs constitute another high-risk group and account for an estimated 15-20% of all reported cases of hepatitis B (4). The ACIP states that HBV vaccine should be given to all susceptible users of illicit injectable drugs and recommends vaccination as soon as possible after the drug use begins. The issues involved in assuring adequate protection of this group are identifying the specific individuals in need of vaccine and identifying these individuals early enough to insure protection prior to infection. The difficulties involved in trying to identify individuals who would prefer to remain hidden are obvious. Furthermore, many illicit drug users are not easily reached through the existing health care system. Currently there are an estimated 3500-4000 drug treatment programs throughout the country (9), but these facilities usually encounter drug users several years after onset of drug abuse, often too late to prevent infection in the majority of individuals. Furthermore, the loss-to-follow-up rate in this population would be high, limiting the number who could be adequately vaccinated with a third dose. Funding for any programs to reach current users or potential users may be difficult to obtain because of social attitudes and the stigma attached to drug abuse. At present, the Army and Veterans Administration are two agencies we are aware of offering free vaccine to drug abusers in rehabilitation programs.

Inmates of Long-term Correctional Facilities

Prevalence surveys have shown that 1-8% of prisoners are carriers, and 10-80% have some HBV marker (1). The real incidence rate of hepatitis B within prisons is not known, but the potential for transmission exists because of homosexual activity, use of illicit drugs and tattooing. The ACIP has stated that because of these practices, prison officials may elect to undertake screening and vaccination programs. Because turnover rates in prisons are high, the time available to vaccinate prisoners may be less than the six months needed to complete a three-dose schedule of vaccine. Current ACIP recommendations are limited to inmates of long-term facilities. As in institutions for the retarded, failure to control the acquisition of new infections in prisons has implications for the rest of society when prisoners are released. Since many prisoners may be held long enough to acquire HBV infection but not long enough to experience clinical infection within prisons, the benefits of protecting short-term prisoners may not be felt in the prison environment. On this account, prison officials may be less willing to secure funding for screening and vaccination, but a few states are examining the cost-effectiveness of vaccination.

Patients in Hemodialysis Units

According to surveillance data collected by CDC's Hepatitis Branch, the incidence of acute hepatitis B has declined among hemodialysis patients from 3% in 1976 to 1% in 1980. The point prevalence of HBsAg positive patients has declined from 7.8% to 3.8% in the same time period (10). Despite the current lower level of risk, this group constitutes a high-risk population, and the ACIP recommends vaccination for all susceptibles in this environment.

What are the chances that this population can be adequately protected against HBV infection? Dialysis patients are well identified in the United States. Currently, expenses for dialysis-associated patients are borne by the Health Care Financing Administration, and authorization to purchase vaccine for dialysis patients is being sought from the federal government. Vaccination in some centers is awaiting the outcome of this request. Vaccine costs will be greater than for healthy individuals since the recommended dosage for hemodialysis patients is twice that for healthy persons. The current costs of controlling HBV infection in this population are enormous, including frequent serological evaluations to monitor the status of every dialysis patient and staff. Since routine screening can be eliminated among all seroconverters, use of the vaccine would be cost-effective on this basis alone. Vaccine efficacy in this patient population may be lower than among healthy persons, so vaccination alone might not achieve complete prevention of HBV infection.

Homosexually Active Males

An estimated 20% of all reported cases of hepatitis B occur among homosexual men (4). The ACIP recommends vaccination for susceptible homosexually active men, regardless of age and duration of homosexual practices, but as soon as possible after homosexual activity begins. Many of the issues to be faced in achieving control of HBV infection among gay men are the same as those to be faced in protecting users of illicit injectable drugs. First, identification of the individuals at risk is complicated by the desire of many to remain anonymous. Currently, an estimated 8% of the United States male population is considered actively homosexual, but only about half of gay men openly and willingly declare their sexual preference (11). Second, many begin their homosexual activity in adolescence, and reaching teenagers or young adults in high schools and colleges prior to exposure will pose special difficulties. In all likelihood some individuals will be accessed through the numerous venereal disease clinics located throughout the United States, but susceptibility to HBV may be very low by the time this population requires treatment for venereal disease. Currently, gay men in the Army and those who qualify for VA assistance may receive hepatitis B vaccine free of charge. Also, gay groups in some large cities are currently planning or conducting screening and vaccination programs, but funding remains an issue.

Household Contacts of HBV Carriers

Household contacts of HBV carriers constitute another high-risk group, with about a third of contacts exhibiting some serologic evidence of infection (12,13). Vaccination is recommended for all susceptible household contacts, regardless of age or relationship to the HBV carrier in the household. The critical difficulty in implementing this recommendation lies in identifying the households at risk, and secondarily in identifying the susceptible contacts. There are no routine screening programs at present in the general population of the United States designed specifically to identify HBV carriers. Perhaps the closest is the routine screening of blood donors which identifies approximately 8000 HBsAg positive individuals per year (14). At the present time, the interest in screening these individuals is to eliminate them from the donor pool, and no special provisions are made to arrange for further testing of the household contacts. Since this group comes from all segments of society and is not part of any captive population, there are no established programs through which they could be served. Vaccination of entire families would be expensive, although screening for spouses at least is likely to be cost-effective, and payment would have to be borne by the individual families or their insurance carriers.

Health Care Workers

Numerous studies have documented an increased prevalence of HBV markers among a variety of health workers (15,16). Depending on the occupational categories studied, health workers have been classified at low or intermediate risk relative to the general population. The ACIP has recommended vaccination for all health workers who are at substantial risk of infection but has not specifically identified each occupational category for whom vaccination is desirable. Perhaps more controversy has surrounded this recommendation than that for any other group, and the issues center around the identification of occupational categories for which vaccination is recommended. The decision has important economic implications for hospitals and is further complicated by the variation in risk which may be exhibited by the same categories of health professionals working in different hospitals. The ACIP recommendation suggests that each hospital or medical group ranks its employees on the basis of risk of exposure. Therein lies a difficulty because many hospitals or medical groups do not have the specific information or, sometimes, as in the case of small hospitals, the expertise to confidently select specific occupational groups as candidates for vaccination. Published data about risks may not always be applicable, but the ACIP has identified some occupational categories as having frequent contact with blood and as being at increased risk. A particularly controversial issue is whether or not to include the housekeeping staff and regular floor nurses as candidates for vaccination. The Veterans Administration has specifically ranked its hospital employees into high, moderate, modest and low risk categories based on the distribution of 258 acute cases it examined over a three-year period (17). The vaccine is being offered free to employees in the high-risk groups. Once subgroups within the medical field have been identified for vaccination, what is the likelihood they will actually be vaccinated? Because of the plasma origins of the vaccine, hypothetical questions have been raised about possible reactions to blood substances or to infectious agents present in donor plasma. This concern has been heightened by the recent recognition of acquired immune deficiency syndrome (AIDS), a clinical entity whose epidemiology suggests an unidentified blood-borne agent as a possible cause of the immunologic defect. AIDS occurs among populations that are sources of HBV positive plasma. An inter-agency group has reviewed the procedures of vaccine preparation and all available data, and has concluded that the vaccine is safe (18). The Centers for Disease Control is monitoring vaccine use to detect any other rare reactions which may be associated with this vaccine. Despite assurances from the inter-agency group, concerns about the safety of HBV vaccine persist, and a "wait and see" attitude may sometimes prevail.

Vaccination coverage among health workers will also depend on the extent to which hospitals and other health agencies assume financial responsibility for protecting their employees. At the present time each institution is developing its own policies and establishing its own list of priority candidates for vaccination with legal issues clearly in mind, and wide variations in coverage can be expected for the immediate future.

LIKELY IMPACT OF CURRENT STRATEGY

Governed by both epidemiologic considerations indicating that overall HBV prevalence is low in the United States, and by the high cost of the current HBV vaccine, a vaccination strategy limited to high-risk groups has been recommended by the ACIP. We have made a quantitative estimate of the anticipated reduction in hepatitis B incidence as a result of the high-risk group strategy. We used reported clinical cases of hepatitis B in the United States and were able to classify 55% of the cases as to possible source of infection (Table 2). We then estimated the proportion of these cases which would be accessible to vaccination. For example, we assumed that only 25% of drug users but 90% of medical/dental personnel would receive vaccine. Except for sexual contacts of acute cases immunized post exposure, we assumed vaccine efficacy at 90% for all groups. We then estimated the proportion of all infections in each high-risk group that would be prevented under these assumptions of coverage and efficacy. Estimates ranged from a low of 23% among drug users to a high of 81% among medical personnel. Overall, we estimated that only 24% of future infections could be prevented. If the 45% of infections with unknown source are derived from chains of infection originating among the known high-risk groups, then this would increase somewhat the estimated percent of infections prevented by a limited vaccination approach.

We can describe two possible scenarios regarding the use of hepatitis vaccine in the near future. With adequately funded programs which seek out high-risk group individuals and vaccinate them free of charge, an anticipated reduction of approximately 25%, as predicted in the model, might be achieved. As minimal as this impact may be, it represents a best-case situation. Without adequately funded programs, which represents the current situation, we foresee that only the most well-informed and most prosperous members of the high-risk groups will be vaccinated. Under this present scenario, an even smaller decrease in hepatitis will follow. The latest available distribution data for HBV vaccine provide further substantiation for this prediction. Through October 1982, 656,000 doses of hepatitis B vaccine have been sold, sufficient to immunize approximately 200,000 persons, or 2% of the 9 million persons estimated by the manufacturer to be in need of vaccine. The majority of all doses has been purchased by hospitals or other agencies, primarily for medical staff, with little evidence of significant vaccine purchases for any other high-risk groups. Since neither of these scenarios appears likely to produce a large reduction in hepatitis B, universal vaccination

Table 2. Source of HBV infection by high-risk groups and projected percent of HBV infection preventable by immunization in the United States[a]

	Sexual contact with acute case	Drug users	Male homosexuals	Medical/ dental	Miscellaneous	Total
Estimated percent of all reported patients	6%	17%	21%	6%	5%	55%
Percent of group accessible to vaccine program	70%	25%	50%	90%	90%	
Efficacy of vaccine	50%	90%	90%	90%	90%	
Proportion of future HBV infections prevented	0.35	0.23	0.45	0.81	0.81	
Percent of infections prevented by limited vaccination U.S.	2.1%	3.9%	9.5%	4.9%	4.0%	24.4%

[a]Source: Sentinel County Data, Hepatitis and Viral Enteritis Division, Centers for Disease Control, Atlanta (4).

during childhood would probably be required, even for a low prevalence area such as the United States, to have any major impact on the overall incidence of hepatitis B infection. At present, we can identify four major constraints to the adoption of a universal vaccination strategy. First is the issue of vaccine cost. Until such time as less expensive HBV vaccines can be developed, mass immunization is not economically feasible. Second is the issue of vaccine safety. Although the vaccine has now been administered safely to a few hundred thousand persons, questions remain about hypothetical risks and about the incidence rate of rare reactions. Until greater experience is accumulated, it is unlikely that the United States population would accept mass immunization for all its children. Third is the issue of the duration of protection and the need for booster doses. If vaccine induced immunity is found to wane and booster doses are required every 5 to 10 years, the cost of a universal program would increase and its effectiveness would be compromised. Fourth, since the bulk of United States infections occur primarily in young adults, universal vaccination in childhood will not produce a significant reduction in HBV infections until the cohort of vaccinated individuals reaches the period of risk 15-20 years later. This anticipated delay in reaping the benefits of universal vaccination would require the use of additional resources to continue limited vaccination of high-risk groups who are in more immediate need of protection or would require mass catch-up programs.

SUMMARY

Based on existing epidemiologic data, recommendations to vaccinate high-risk groups have been formulated in the United States. While such a limited approach has obvious economic and practical advantages in that the number of individuals to be vaccinated is less than in a more comprehensive approach, the limited strategy poses its own special difficulties. Among these are the need to specifically identify the individuals in need of vaccine, to vaccinate them prior to exposure, and to achieve adequate coverage. With more adequately funded programs, a decrease of approximately 25% in hepatitis B incidence might be anticipated. Without additional funding to actively reach high-risk groups, as in the present situation, a more modest decline is expected. Existing constraints such as cost, safety concerns and an unknown duration of vaccine induced immunity preclude the adoption of a more universal approach at the present time. The effective control of HBV infection is likely to remain as a public health challenge for the foreseeable future, perhaps no less challenging than the scientific achievements which have produced the vaccine.

REFERENCES

1. MMWR. (1982). 31:317-322; 327-328.
2. Sobeslavsky, O. (1978). HBV as a global problem, in Viral Hepatitis, G. N. Vyas, S. N. Cohen and R. Schmid, eds., Franklin Institute Press, Philadelphia, pp. 347-355.
3. CDC. (June 1982). Hepatitis Surveillance Report #48.
4. Hadler, S. Personal communication, CDC, Atlanta.
5. Department of State, Bureau of Consular Affairs, Statistics on Immigrant Visa Issuances, FY80-81.
6. Department of State, Refugee Program, Indochinese Refugee Program Summary, period ending November 30, 1982.
7. California Morbidity Supplement, #9, March 12, 1982.
8. Schreeder, M. T., Moser, M. R., Bender, T. R., Sheller, M. J., Murphy, B. L., McMahon, B. J., Heyward, W. L., Hall, D. B., Berquist, K. R., and Maynard, J. G. Prevalence of hepatitis B in Alaska Eskimos. Am. J. Epid. (in press).
9. Ginzberg, H. Personal communication, National Institute on Drug Abuse, Washington, D.C.
10. Alter, M. Personal communication, CDC, Atlanta.
11. Darrow, W. Personal communication, CDC, Atlanta.
12. Szmuness, W., Harley, E. J., and Prince, A. M. (1975). Intrafamilial spread of asymptomatic hepatitis B. Am. J. Med. Sci. 270:293-304.
13. Bernier, R. H., Sampliner, R., Gerety, R., Tabor, E., Hamilton, F., and Nathanson, N. (1982). Infection in households of chronic carriers of hepatitis B surface antigen. Am. J. Epid. 116:199-211.
14. Dodd, R. Y., Nath, N., Bastiaans, M. D., and Barker, L. F. (1982). Hepatitis-associated markers in the American Red Cross donor population: Ten years experience, in Viral Hepatitis, 1981 International Symposium, W. Szmuness, H. J. Alter and J. E. Maynard, eds., Franklin Institute Press, Philadelphia, pp. 145-155.
15. Maynard, J. E. (1981). Nosocomial viral hepatitis. Am. J. Med. 70:439-444.
16. Dienstag, J. L. and Ryan, D. M. (1982). Occupational exposure to hepatitis B virus in hospital personnel: Infection or immunization. Am. J. Epid. 115:26-39.
17. Mather, S. Personal communication, Veterans Administration, Washington, D.C.
18. MMWR. (1982). 31:465-467.
19. Williamson, T. Personal communication, Merck Sharpe & Dohme, West Point, PA.

NEWLY LICENSED HEPATITIS B VACCINE

Robert J. Gerety

Office of Biologics
National Center for Drugs and Biologics
Food and Drug Administration
Bethesda, Maryland 20205

The recently licensed subunit hepatitis B vaccine (HEPTAVAX-B) is unique among vaccines in that it is manufactured solely from human plasma obtained from asymptomatic individuals with chronic hepatitis B. Plasma from donors selected for manufacturing this vaccine contains high concentrations of noninfectious hepatitis B surface antigen (HBsAg) particles and lower concentrations of infectious hepatitis B virus (HBV). The ratio of HBsAg to HBV can be as great as 10,000:1 (1). From the starting plasma, 22 nm spherical HBsAg particles are separated from HBV by ultracentrifugation. The 22 nm particles are then further purified by digestion with pepsin followed by the addition of 8M urea. Finally, the vaccine is treated with formaldehyde.

VACCINE PLASMA DONORS

Vaccine plasma donors are selected for their high titers of HBsAg. By federal regulation, they must be asymptomatic and in apparent good health (2). They must meet all federal requirements for acceptable plasmapheresis donors, except that their serum aminotransferase activity may exceed the level permitted normal donors (but it must be stable) (3). Each donor provides a complete history, receives a complete physical examination, and undergoes laboratory tests before his or her first plasma donation and must maintain normal levels of hemoglobin and serum protein and a normal hematocrit value throughout the course of plasma donations (2,3). In the general population, there are fewer than two acceptable (high titer) vaccine plasma donors among each 10,000 individuals.

Among certain high-risk populations for hepatitis B, e.g. sexually active male homosexuals, there are approximately 80 acceptable vaccine plasma donors per 10,000 individuals.

VACCINE MANUFACTURE

To isolate the 22 nm noninfectious HBsAg particles, acceptable plasma pools are treated with ammonium sulfate (concentration step) followed by isopyknic ultracentrifugation in sodium bromide and rate zonal ultracentrifugation in sucrose. The purified 22 nm particles are then treated with pepsin (1 mg/L, pH 2.1 at 37°C for 18 hours) to digest residual plasma antigens, with 8M urea for four hours followed by dialysis, and with 1:4000 formaldehyde solution at 37°C for 72 hours (4).

VACCINE SAFETY

Ultracentrifugation of plasma to isolate noninfectious HBsAg particles has been shown to be capable of removing 10,000 infectious doses of HBV (1). Each of the three steps routinely applied during vaccine manufacture (pepsin, urea, and formaldehyde) has been individually shown to inactivate 100,000 infectious doses of HBV per milliliter by chimpanzee inoculation studies (5).

The procedures used in the manufacture of hepatitis B vaccine are also effective in inactivating viruses from every known group. The pepsin treatment has been shown to inactivate completely rhabdoviruses represented by vesicular stomatitis virus, poxviruses represented by vaccinia, togaviruses represented by sindbis, herpesviruses represented by herpes simplex (type 1), coronaviruses represented by infectious bronchitis virus, and reovirus (6,7). The urea treatment, in addition to inactivating all of the previously mentioned viruses, also has been shown to inactivate completely myxoviruses represented by Newcastle disease virus and picornaviruses represented by mengovirus (6,7); lesser concentrations of urea have been shown to inactivate viruses in the slow virus category represented by the scrapie agent (8). Formalin inactivates a wide variety of viruses, including parvoviruses, and has been recently shown in two separate studies to inactivate agents of human non-A, non-B hepatitis (9,10).

VACCINE TESTING AND EXPERIENCE

Each lot of licensed hepatitis B vaccine is injected into suckling and adult mice, guinea pigs, the allantoic and yolk sac of embryonated eggs, and into animal (Vero cells) and human-derived cell cultures (WI-38 cells) to confirm the absence of infectious agents. In addition, a total of 22 doses of each vaccine lot are injected intravenously (i.v.) into chimpanzees that are examined for six months with weekly serum analyses and monthly liver biopsies. A total of more than 15 lots of the newly licensed hepatitis B vaccine has been tested in this manner, with no evidence of residual infectivity for hepatitis B or any other viral agent.

During clinical trials of the licensed vaccine involving 1900 vaccinees (11,12), only minor immediate reactions occurred. Complaints primarily concerned soreness at the injection site. To date, more than 19,000 persons have received this vaccine, including 8941 health care workers and 5985 healthy adults, children and infants who were not at high risk for acquiring hepatitis B; no serious reactions have been proven to be vaccine related so far (13).

ACQUIRED IMMUNE DEFICIENCY SYNDROME

The recognition of the acquired immune deficiency syndrome (AIDS) in previously healthy homosexual males, in Haitian immigrants to the United States, in addicts using drugs i.v., and in 84 heterosexual men and 34 women has led to speculation that an infectious agent may be responsible for this disorder. Acquired immune deficiency syndrome is recognized by the occurrence of either a life-threatening opportunistic infection (such as _Pneumocystis carinii_) or Kaposi's sarcoma in a person younger than 60 years who has no underlying immunosuppressive disease and has not received immunosuppressive therapy (14). Mortality from AIDS has ranged from 15% for those with Kaposi's sarcoma to 60% for those with _P. carinii_ pneumonia (14)--far higher than the mortality seen from these disorders in patients without AIDS. Many patients with AIDS had a prodrome lasting weeks to months characterized by weight loss, lymphadenopathy, fever and diarrhea. Although this prodrome was more common in patients with _P. carinii_ (14), it was by no means universal among patients with AIDS. It is not clear whether this prodromal lymphadenopathy reflects an immune dysfuncton predating the recognition of AIDS. In this regard, Kaposi's sarcoma in patients without AIDS may be associated with lymphoreticular neoplasms (15), suggesting that an immune defect may precede all cases of Kaposi's sarcoma.

ETIOLOGY OF AIDS

The etiology of AIDS is unknown. Suggested causes include single or repeated exposures to an unknown immunosuppressive agent or to a known agent, e.g. cytomegalovirus (14,16). Other suggested causes include exposures to antigenic substances, opiates (17), nitrite inhalants (14), or chemotherapeutic agents to which homosexual males may be exposed during treatments for various disorders. The possibility of herpes simplex virus infection being a predisposing factor has not yet been discussed, although there is a high prevalence of infections by this agent among homosexual males and among patients with AIDS (18,19).

The recognition of cases of AIDS in heterosexual hemophilic patients, although in no way proved to be related to the life-saving blood-derived clotting factor concentrates that they receive, has raised questions regarding the safety of plasma donated for hepatitis B vaccine manufacture by persons with chronic hepatitis B who may have unrecognized or early AIDS. It is important that none of the hemophilic patients with AIDS described so far had received clotting factor concentrate from a common lot of product, despite having required frequent treatments (20) and despite the fact that these same lots had been infused into several hundred other hemophiliacs without resulting in AIDS. Also, a relatively small number of human viruses have been documented to be transmitted by blood (21), and, in some cases, these are "intracellular" viruses, e.g. cytomegalovirus, that are not thought to be transmitted by plasma. Active surveillance among hemophiliacs will continue to identify any additional AIDS cases in this group.

RISK-BENEFIT CONSIDERATIONS AND VACCINE USE POLICY

Although not convincingly documented to be transmitted by blood, AIDS does occur among populations that donate vaccine plasma. Since AIDS occurs among certain high-risk groups receiving the vaccine, it is inevitable that cases of AIDS will occur in vaccine recipients unrelated to the vaccine itself. There is little doubt that the licensed vaccine is highly effective in preventing hepatitis B (22). There are about 200,000 new cases of hepatitis B each year in the United States, 10,000 of these patients require hospitalization, up to 1000 die with fulminant hepatitis, and 10,000 to 20,000 experience chronic hepatitis B. It is estimated that half of these cases of hepatitis B can be prevented by immunizing all persons at high risk for hepatitis B with the licensed vaccine (22,23).

Despite the fact that a great body of data has not yet been accumulated from long-term evaluation of hepatitis B vaccine recipients, it is clear at this point that the known risk of hepatitis B for persons in high-risk groups far exceeds the risks of vaccine-induced infection by a theoretical transmissible agent that would have to survive the purification and inactivation procedures applied to the licensed hepatitis B vaccine. The recommendations of the Immunization Practices Advisory Committee

(23) have recently been reaffirmed (13); all persons at high risk for hepatitis B should receive hepatitis B vaccine.

REFERENCES

1. Gerety, R. J., Tabor, E., Purcell, R. H., et al. (1979). Summary of an international workshop on hepatitis B vaccines. J. Infect. Dis. 140:642-648.
2. Code of Federal Regulations, part 21, section 640. Government Printing Office, 1979.
3. FDA Guidelines: Clarification of Guidelines for Plasmapheresis of HBsAg Reactive Donors. Bethesda, MD, Food and Drug Administration, March 1981.
4. Hilleman, M. R., Buynak, E. B., McAleer, W. M., et al. (1982). Hepatitis A and hepatitis B vaccines, in Viral Hepatitis, 1981 International Symposium, W. Szmuness, H. J. Alter and J. E. Maynard, eds., Franklin Institute Press, Philadelphia, pp. 385-397.
5. Tabor, E., Buynak, E., Smallwood, L. A., et al. Inactivation of hepatitis B virus by three methods: Treatment with pepsin, urea, or formalin. J. Med. Virol., in press.
6. Buynak, E. B., Roehm, R. R., Tytell, A. A., et al. (1976). Development and chimpanzee testing of a vaccine against human hepatitis B. Proc. Soc. Exp. Biol. Med. 151:694-700.
7. Hilleman, M. R., Buynak, E. B., Roehm, R. R., et al. (1975). Purified and inactivated human hepatitis B vaccine: Progress report. Am. J. Med. Sci. 270:401-404.
8. Hunter, G. D., Gibbons, R. A., Kimberlin, R. H., et al. (1969). Further studies of the infectivity and stability of extracts and homogenates derived from scrapie affected mouse brains. J. Comp. Pathol. 79: 101-108.
9. Tabor, E. and Gerety, R. J. (1980). Inactivation of an agent of human non-A, non-B hepatitis by formalin. J. Infect. Dis. 142:767-770.
10. Yoshizawa, H., Itoh, Y., Iwakiri, S., et al. (1982). Non-A, non-B (type 1) hepatitis agent capable of inducing tubular ultrastructures in the hepatocyte cytoplasm of chimpanzees: Inactivation by formalin and heat. Gastroenterology 82:502-506.
11. Szmuness, W., Stevens, C. E., Harley, E. J., et al. (1980). Hepatitis B vaccine: Demonstration of efficacy in a controlled clinical trial in a high-risk population in the United States. N. Engl. J. Med. 303:833-841.
12. Francis, D. P., Hadler, S. C., Thompson, S. E., et al. (1982). The prevention of hepatitis B with vaccine: Report of the CDC multi-center efficacy trial among homosexual men. Ann. Intern. Med. 97:362-366.

13. Hepatitis B virus vaccine safety: Report on an inter-agency group. (1982). MMWR 31:465-467.
14. Centers for Disease Control Special Report: Epidemiologic aspects of the current outbreak of Kaposi's sarcoma and opportunistic infection. (1982). N. Engl. J. Med. 306:248-252.
15. Safai, B. and Good, R. A. (1981). Kaposi's sarcoma: A review and recent developments. CA 31:2-12.
16. Drew, W. L., Conant, M. A., Miner, R. C., et al. (1982). Cytomegalovirus and Kaposi's sarcoma in young homosexual men. Lancet 2:125-127.
17. Masur, H., Michelis, M. A., Greene, J. B., et al. (1981). An outbreak of community-acquired _Pneumocystis carinii_ pneumonia: Initial manifestation of cellular immune dysfunction. N. Engl. J. Med. 305:1431-1438.
18. Friedman-Kien, A. E., Laubenstein, L. J., Rubinstein, P., et al. (1982). Disseminated Kaposi's sarcoma in homosexual men. Ann. Intern. Med. 96:693-700.
19. Mildvan, D., Mathur, U., Enlow, R. W., et al. (1982). Opportunistic infection and immune deficiency in homosexual men. Ann. Intern. Med. 96:700-704.
20. _Pneumocystis carinii_ pneumonia among persons with hemophilia A. (1982). MMWR 31:365-367.
21. Tabor, E. (1982). Infectious Complications of Blood Transfusion, Academic Press Inc., New York.
22. Krugman, S. (1982). The newly licensed hepatitis B vaccine: Characteristics and indications for use. JAMA 247:2012-2015.
23. Inactivated hepatitis B virus vaccine. (1982). MMWR 31: 318-328.

CHARACTERIZATION AND MAPPING OF VIRAL AND PUTATIVE VIRAL-CELLULAR TRANSCRIPTS IN A HEPATITIS B VIRUS INFECTED HUMAN HEPATOMA CELL LINE AND IN CHIMPANZEE CARRIER LIVER

Prasanta R. Chakraborty[1], Nelson Ruiz-Opazo[2] and David A. Shafritz[3]

[1]Department of Microbiology and Immunology, The Medical College of Pennsylvania, Philadelphia, PA

[2]Department of Pediatrics/Cardiology, Children's Hospital Medical Center, Harvard Medical School Boston, MA

[3]Departments of Medicine/Cell Biology, Albert Einstein College of Medicine, Bronx, New York

INTRODUCTION

Since attempts to propagate hepatitis B virus (HBV) in tissue culture have generally been unsuccessful, recent studies on the mechanism for viral morphogenesis have focused on expression of the HBV genome in HBV infected human hepatoma cell lines (1,2,3,4) and *in vitro* transfected rodent cells (5). The human hepatoma cell lines PLC/PRF/5 and HEP-3B synthesize and secrete 22-nanometer particles with immunochemical characteristics of HBsAg and contain HBV DNA exclusively in integrated form (1,2,3,6,7). Initial reports of viral RNAs expressed in PLC/PRF/5 cells have given conflicting results. Marion et al. (1) demonstrated sequences hybridizing with the whole genome of HBV by solution hybridization, whereas Edman et al. (3) found only one transcript of 2000 nucleotides, which hybridized to the S region (HBsAg). However, in L cells transformed with cloned HBV DNA and in PLC/PRF/5 cells, Pourcel et al. (8) recently detected a 2.3 kb HBV-specific poly $(A)^+$ RNA. This RNA hybridized to the pre-S region, to the S region, and to a 1100 bp region downstream from the S gene coding sequence; and it was proposed that the sequence 5'-TATATAA starting at position 2776 could be related to the S gene promoter (8).

The present study shows that PLC/PRF/5 cells express 5 cytoplasmic poly A^+ RNAs containing different portions of the HBV genome. Two of these cytoplasmic poly A^+ RNAs (17S and 19S) contain HBsAg gene sequences. RNA mapping studies suggest that a viral RNA initiation site (promoter) is utilized for transcription of both of these RNAs and that a splicing event may be involved in nuclear maturation of these viral RNAs. These studies suggest further that the 19S poly A^+ RNA represents a cotranscript of viral and cellular sequences. Two higher molecular weight poly A^+ RNAs (21S and 26S) contain a small portion of the HBV genome in the HBcAg gene region and might also represent cotranscripts of viral and cellular sequences. A 37S poly A^+ RNA contains viral sequences from all regions of the HBV genome. Parallel studies with RNA isolated from the livers of chronic HBV chimpanzee carriers showed three poly A^+ RNAs of 24S, 19S and 14S. The chimpanzee 24S poly A^+ viral RNA represents a complete transcript of the HBV genome, and the chimpanzee 19S viral RNA is distinct from the 19S poly A^+ RNA found in PLC/PRF/5 cells, since it contains both HBsAg and HBcAg gene sequences. The finding of multiple HBV RNA transcripts in PLC/PRF/5 cells suggests that several viral transcriptional units may be active in this system. Based on the findings in PLC/PRF/5 cells vs. chimpanzee liver, these studies indicate that the nature of the viral transcripts depends on the physical state of the viral DNA (i.e., whether this DNA is integrated versus extrachromosomal).

MATERIALS AND METHODS

Preparation of RNA from PLC/PRF/5 cells

PLC/PRF/5 cells were grown to confluence in monolayer culture using Eagle's minimal essential medium supplemented with glutamine, non-essential amino acids, 10% heat-inactivated fetal calf serum, penicillin, streptomycin and fungizone as previously reported (9). Trypsinized cells were lysed with detergent essentially as described by Innis and Miller (10), and total cytoplasmic RNA was isolated from the post-nuclear supernatant fraction by SDS/phenol/chloroform/isoamyl alcohol extraction at 65°C, followed by ethanol precipitation. Cytoplasmic poly A^+ RNA was recovered by oligo (dT)-cellulose column chromatography (11).

Preparation of RNA from infected chimpanzee liver (CH30)

Frozen liver tissue ($\sim$ 3 gm) was homogenized in 5 volumes of buffer containing 50 mM Tris HCl, pH 7.5, 50 mM NaCl, 5 mM $MgCl_2$, 0.14 M sucrose, 1% deoxycholic acid, and 2% NP-40. After 7-8 strokes in a Potter teflon homogenizer, the suspension was centrifuged at 5000 x g for 10 min. The nuclei and cellular debris were discarded. The supernatant was adjusted to 1% SDS and subsequently extracted twice with a mixture of phenol:chloroform:isoamyl alcohol, 50:50:1 (v/v/v), and then twice with ether. Finally, total cytoplasmic RNA was ethanol precipitated by addition of 0.2 M sodium acetate, pH 5.5 and 3 volumes of ethanol. Cytoplasmic poly A^+ enriched RNA was isolated by oligo dT-cellulose column chromatography (11).

Purification of restriction fragments from cloned HBV DNA

The subgenomic fragments depicted in Figure 1 were isolated by restriction endonuclease digestion of purified, cloned HBV DNA, followed by agarose gel electrophoresis and electroelution of DNA bands from the gel. Each fragment was reisolated through a second agarose gel electrophoresis and subsequent elution. The twice-purified fragments showed no cross-contamination, as judged by hybridization to a high specific activity [^{32}P]-labelled HBV DNA probe (2-4 x 10^8 cpm/ug).

[^{32}P]-labelling of HBV DNA probes

Purified HBV DNA fragments were [^{32}P]-labelled by nick translation, essentially as described previously (2). Both [α^{32}P]-dATP and [α^{32}P]-dCTP (spec. act. 300-800 Ci/mmole) were included to obtain a specific activity of 0.3-0.5 x 10^8 cpm/ug for subgenomic fragments.

Gel electrophoresis, blotting and hybridization procedures

RNA was denatured with glyoxal essentially as described by McMaster and Carmichael (12). RNA samples were resuspended in 40 ul of a denaturing buffer containing 80% DMSO (dimethylsulfoxide), 1.5 M glyoxal and 10 mM Na_3PO_4, pH 6.8. For each use, glyoxal was freshly deionized with a mixed bed resin, AG 501-X8 (D), from Bio-Rad Laboratories. Samples were incubated at 50°C for 1 hour, followed by addition of a one-tenth volume of stop solution (1% SDS, 0.1 bromophenol blue, 100 mM EDTA, 50% glyercol) for subsequent gel electrophoresis. RNA samples were resolved by electrophoresis through 1% agarose submerged slab gels (15 x 15 x 0.5 cm). Electrophoresis was carried out in the buffer of Dingman and Peacock (13) at 120 volts for 3-4 hours.

Agarose gel slabs containing RNA after electrophoresis were soaked in a dilute solution of alkali (0.05 N NaOH) for 20 min. The transfer of nucleic acids to DBM paper was performed according to the procedure of Alwine, Kemp and Stark (14). After transfer, DBM papers were rinsed with 2 x SSC, prehybridized and hybridized as indicated above. After hybridization, the DBM papers were washed three times (15 min. each) with 2 x SSC, 0.1% SDS at 45°C and three times (15 min. each) with 0.1 x SSC, 0.1% SDS at 45°C. Finally, the papers were dried and autoradiographed.

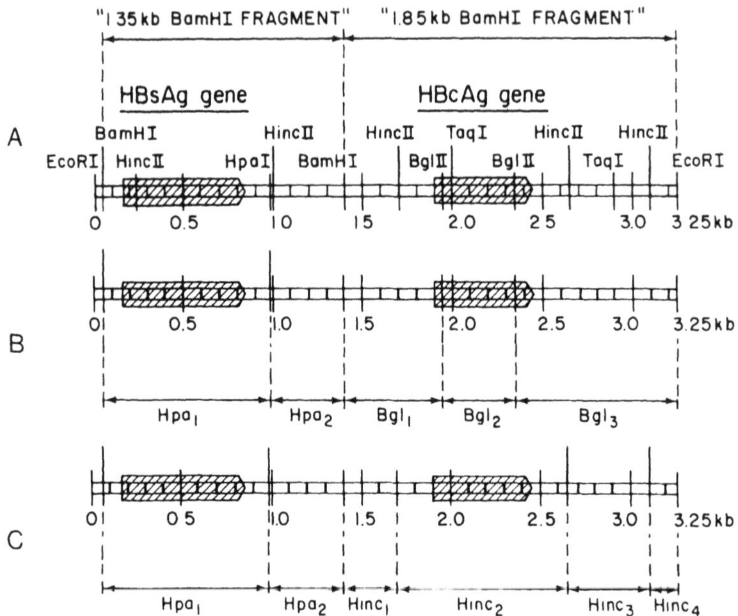

Figure 1. Restriction map of cloned HBV DNA/adw and schematic diagrams of restriction endonuclease sites relevant for isolation of Hpa_1, Bgl II and Hinc II subgenomic HBV DNA probes. The position of cleavage sites for some restriction endonucleases are indicated on the EcoRI linearized HBV DNA map. The location and direction of gene coding sequences for HBsAg and HBcAg are shown with hatched arrows (1A). The lengths and the position of the subgenomic DNA fragments are indicated at the bottom of the scheme. The purified 1.35 kb Bam HI fragment produced Hpa_1 and Hpa_2 fragments upon digestion with restriction endonuclease Hpa_1 (1B). Bgl II digestion of the purified 1.85 kb Bam HI fragment generated Bgl_1, Bgl_2 and Bgl_3 subgenomic DNA fragments (1B). Similarly, Hinc II digestion of the purified 1.85 kb Bam HI fragment generated $Hinc_1$, $Hinc_2$, $Hinc_3$ and $Hinc_4$ subgenomic DNA fragments (1C). All of these fragments were twice agarose gel purified as described in MATERIALS AND METHODS.

5'-end-labelling of PLC/PRF/5 poly A^+ RNA

The 5'-terminus of most eukaryotic and viral mRNA contains a "cap" structure comprised of 7-methyl-guanosine in 5' to 5' triphosphate linkage to the second nucleotide which is usually 6-methyl-A or G [for review, see Shatkin (15)]. For *in vitro*

[^{32}P]-labelling of the 5' terminus of eukaryotic mRNA, the 7-methyl-guanosine and 5'-phosphates moieties of the "cap" were removed by tobacco acid pyrophosphatase, TAP (16), followed by dephosphorylation with bacterial alkaline phosphatase (BAP) to generate a free 5'-OH group and 5'-end [^{32}P]-labelling with [^{32}P] ATP and polynucleotide kinase. Thirty ug of PLC/PRF/5 poly A$^+$ RNA was treated with 10-15 units of TAP (Bethesda Research Laboratories, Inc., Bethesda, MD) in a 50 ul reaction volume containing 50 mM Na acetate, pH 5.0, 1 mM Na_2 EDTA and 10 mM 2-mercaptoethanol. Incubation was performed at 37°C for 30 min. RNA was isolated by hot SDS-phenol extraction, followed by ethanol precipitation. TAP treated RNA was dephosphorylated by adding 15-20 units of BAP in 10 mM Tris HCl, pH 8.0, at 37°C for 60 min. After the reaction was completed, BAP treated RNA was isolated by hot SDS-phenol extraction. This extraction procedure was repeated three times to ensure complete removal of BAP. RNA was then precipitated with ethanol. The RNA pellet was resuspended in 20 ul of 10 mM Tris HCl, pH 7.5, 1 mM spermidine and heated to 50°C for 3 min., followed by chilling in ice water. For the kinase reaction, the dephosphorylated RNA (30 ug) was incubated in a 50 ul reaction volume containing 50 mM Tris HCl, pH 7.5, 10 mM $MgCl_2$, 4 mM DTT, 5% glycerol, 1 uM [^{32}P] ATP (3000-5000 Ci/mmole, Amersham, Searle) and 20-30 units of T4 polynucleotide kinase (PL Biochemicals) for 30 min. at 37°C. The reaction was stopped by addition of SDS and EDTA to a final concentration of 0.1% and 20 mM, respectively. 5'-[^{32}P]-end-labelled RNA was separated from unreacted [^{32}P] ATP by passage through a Biogel P-60 column (10 ml bed volume) and collection of radioactive material eluting in the excluded volume of the column.

Blotting and hybridization procedures for 5'-poly A$^+$ RNA protection assay

HBV DNA was depurinated, nicked and transferred to diazobenzloxymethyl (DBM) paper (Schleicher and Schuell), according to the procedure of Alwine, Kemp and Stark (14). After transfer, the paper was rinsed with 2 x SSC and prehybridized at 42°C for 16-24 hours in 50% formamide, 5 x SSC, 10 x Denhardt's solution (0.2% bovine serum albumin, 0.2% polyvinyl pyrrolidone and 0.2% Ficoll), 25 mM Na_3PO_4 (pH 6.5), 1% glycine, 0.2% SDS and 0.5-1.0 mg/ml denatured calf thymus DNA. The DBM paper was subsequently hybridized with 40 x 10^6 cpm of [^{32}P]-5'-end-labelled PLC/PRF/5 poly A$^+$ RNA (specific activity of 2 x 10^6 cpm/ug RNA) for 18-24 hours at 42°C in 4 ml of a solution containing 50% formamide, 5 x SSC, 1 x Denhardt's solution, 25 mM Na_3PO_4 (pH 6.5), 0.2% SDS, 100 ug/ml denatured calf thymus DNA and 10% dextran sulfate. After hybridization, the DBM paper was washed three times (15 min. each) with 2 x SSC (SSC is a solution containing 0.15 M NaCl - 0.015 M Na citrate, pH 7.0) at 45°C, and three times (15 min. each) with 10 mM Tris HCl, pH 7.5, 1 mM EDTA, 0.1 M NaCl. Subsequently, the paper

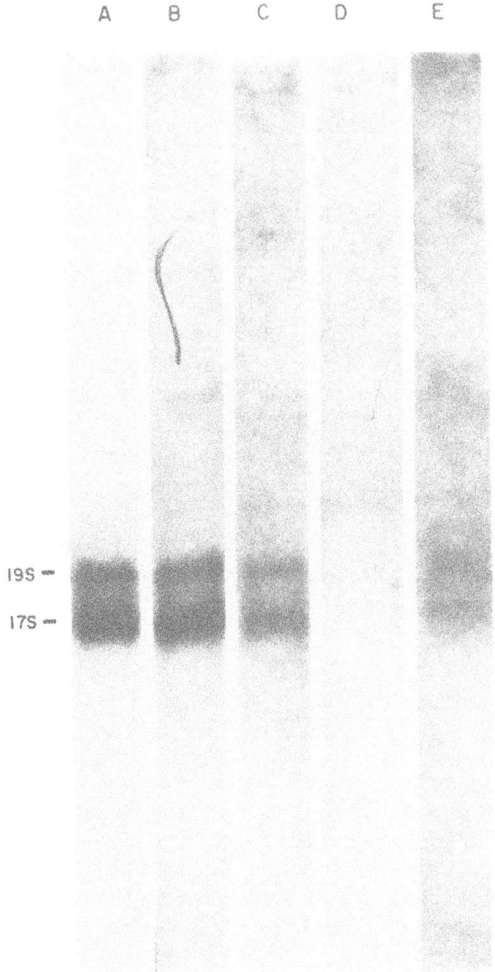
A
B
C
D
E
19S
17S

was treated with 50 ug/ml of pancreatic RNAse in buffer (10 mM Tris, HCl, pH 7.5, 1 mM EDTA, 0.1 M NaCl) at 37°C for 30 min. Finally, the paper was washed three times with 0.1 x SSC, dried and autoradiographed, using Picker Max 3 intensifying screens at -70°C.

RESULTS

Identification and mapping of two cytoplasmic poly A^+ RNAs containing HBsAg gene sequence in the PLC/PRF/5 cell line

Previously, we described the presence of two major viral $poly^+$ RNAs in PLC/PRF/5 cells (2), originally reported as 2.5 and 3.05 kb, respectively. Subsequently, experiments have shown that these RNAs are 17S (1800N) and 19S (2400N), respectively. In order to analyze the HBV sequences represented in these 17S and 19S poly A^+ RNAs, "Northern" blots of total cytoplasmic poly A^+ RNA were performed using specific probes to different portions of the HBV genome. Figure 1 shows a partial restriction map of cloned HBV DNA and the specific fragments used as [^{32}P] labelled probes. Four of the 5 fragments depicted in Figure 1B showed various degrees of hybridization to 17S and 19S poly A^+ RNAs (Figure 2). The Bgl_2 fragment (Figure 2, lane D), which contains HBcAg gene coding sequence, showed no hybridization. Although all 5 probes were labelled to the same specific activity (2 x 10^7 cpm/ug), the Hpa_1 probe (lane A) (950 bp), containing HBsAg gene coding sequences, revealed the strongest hybridization. The Hpa_2

←

Figure 2. Identification of specific HBV DNA sequences in 17S and 19S poly A^+ RNA isolated from PLC/PRF/5 cells using Hpa I and Bgl II HBV DNA fragments as probes. Isolation of poly A^+ RNA from PLC/PRF/5 cells, glyoxal denaturation, electrophoresis, blotting of the RNA to DBM paper and hybridizations were performed as mentioned in MATERIALS AND METHODS. Three μg of poly A^+ RNA were applied to lanes A, B, C, D and E, respectively. Lane A was hybridized to [^{32}P]-labelled Hpa_1 HBV DNA probe, lane B to [^{32}P]-labelled Hpa_2 HBV DNA probe, lane C to [^{32}P]-labelled Bgl_1 HBV DNA probe, lane D to [^{32}P]-labelled Bgl_2 HBV DNA probe, lane E to [^{32}P]-labelled Bgl_3 HBV DNA probe. (For map location of the different restriction fragments, see Figure 1B). The specific activity of these probes was 2 x 10^7 cpm/μg. The autoradiogram for lanes B to E was developed after 8 days of exposure. Lane A was developed after 24 hours of exposure. Apparent molecular weights of 17S and 19S (on the right) were determined using 18S, 25S and 28S ribosomal RNA as markers.

($\sim$400 bp) and Bgl_1 ($\sim$550 bp) HBV DNA probes (lanes B and C) showed similar hybridization, but lower activity than that obtained with the Hpa_1 probe. When the Bgl_3 fragment probe ($\sim$900 bp) was used (lane E), both poly A^+ RNAs showed very low, but detectable, hybridization. The difference in hybridization intensity with these various subgenomic HBV probes was not due to different amounts of poly A^+ RNA loaded onto the gel or to cross contamination of DNA fragments during their purification. Since "nick translation" produces uniformly but randomly labelled DNA molecules, the hybridization intensities reflect the extent of sequence homology between the RNA molecules and the DNA probes used for their detection.

The above studies showed that 17S and 19S poly A^+ RNAs contain HBsAg gene coding sequences but not HBcAg gene sequences (as detected by the Hpa_1 and Bgl_2 probes, respectively). Since the direction of HBsAg gene transcription in relation to the restriction map is known (3,17), hybridization with the Hpa_2 and Bgl_1 probes demonstrates that these two poly A^+ RNAs have 3' sequences extending into the Bgl_1 region of the HBV genome. A small region of the Bgl_3 fragment ($\sim$100 bp) located upstream from HBsAg gene coding sequence is also present in both poly A^+ RNAs (Figure 2, lane E).

In order to map these two poly A^+ RNAs more precisely with respect to their 5' and 3' boundaries, we performed additional experiments using a different set of subgenomic fragment probes (Figure 1C). These probes (Hinc II) permitted finer distinction of sequences in the pre-S region of the HBV genome. As shown in Figure 3, both the Hpa_1 (HBsAg gene coding sequence) and Hpa_2 fragments again hybridized strongly to 17S and 19S poly A^+ RNAs (lanes A and B, respectively). In contrast, the $Hinc_2$ (HBcAg gene sequence) and $Hinc_4$ fragments did not hybridize to these poly A^+ RNAs (lanes D and F, respectively). However, two other Hinc II fragments, $Hinc_1$ ($\sim$300 bp) and $Hinc_3$ ($\sim$450 bp), hybridized to both 17S and 19S RNA molecules (lanes C and E, respectively). It should be noted that although the $Hinc_3$ probe is larger than the $Hinc_1$ probe, it developed a significantly lower hybridization signal. This suggests that only a small portion of DNA sequence in the $Hinc_3$ fragment (see Figure 1C) is represented in both poly A^+ RNAs and corroborates results obtained with the Bgl_3 probe (c.f. Figures 1B and 2E).

The above experiments were performed with the same amount of poly A^+ RNA per lane of the gel and with HBV DNA probes of similar specific activity (2-3 x 10^7 cpm/ug). Therefore, the differential hybridization signals obtained in Figure 3, lanes C ($Hinc_1$ probe) and E ($Hinc_3$ probe) reflect the relative amount of sequences of each fragment represented in 17S and 19S poly A^+ RNAs. Other hybridization bands containing HBV sequences were also detected (Figure 3). These bands are in the high molecular weight region

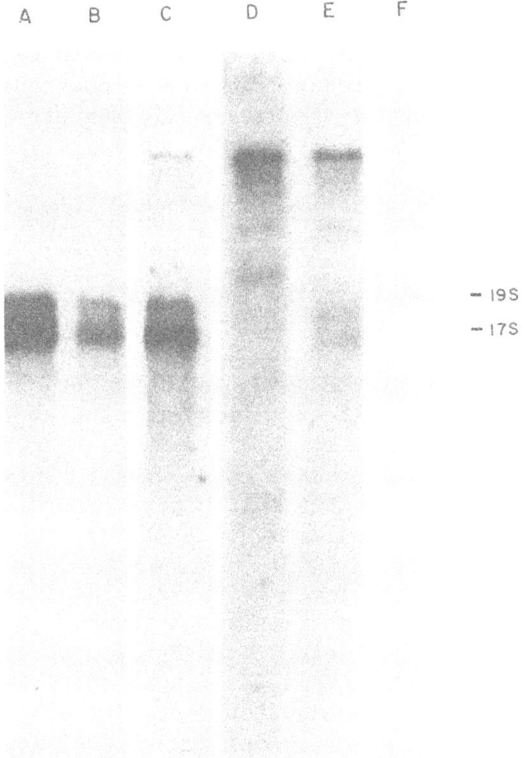

Figure 3. Identification of specific HBV DNA sequences in 17S and 19S poly A^+ RNA isolated from PLC/PRF/5 cells using Hpa I and Hinc II fragments of cloned HBV DNA as probes. Isolation of poly A^+ RNA from PLC/PRF/5 cells, glyoxal denaturation, electrophoresis, blotting of the RNA to DBM paper, and hybridizations were performed as mentioned in MATERIALS AND METHODS. Four μg of poly A^+ RNA were applied to lane A, B, C, D, E and F respectively. Separately, lane A was hybridized with [^{32}P]-labelled Hpa_1 HBV DNA probe, lane B with [^{32}P]-labelled Hpa_2 HBV DNA probe, lane C with [^{32}P]-labelled $Hinc_1$ HBV DNA probe, lane D with [^{32}P]-labelled $Hinc_2$ HBV DNA probe, lane E with [^{32}P]-labelled $Hinc_3$ HBV DNA probe and lane F with [^{32}P]-labelled $Hinc_4$ HBV DNA probe. (See Figure 1C for map location of the different restriction fragments.) The specific activities of these probes were 2-3 x 10^7 cpm/μg. The autoradiogram for lanes A and B was developed after 20 hours of exposure and those for lane C, D, E and F were developed after 5 days of exposure. Apparent molecular weights of 17S and 19S (shown on the right) were determined by using 18S, 25S and 28S ribosomal RNA as markers.

(21S, 26S and 37S) and are in much lower abundance as compared to 17S and 19S poly A^+ RNAs. The 21S and 26S poly A^+ RNAs hybridized to only small regions of the HBV genome containing primarily HBcAg gene sequences. On longer exposure of the autoradiogram, the 37S poly A^+ RNA was present on all lanes and, therefore, hybridized with the entire complement of the HBV genome.

Identification of putative in vivo RNA initiation site(s) for the 17S and 19S poly A^+ RNAs of PLC/PRF/5 cells

The above studies suggested that the *in vivo* RNA initiation site for transcription of the most abundant PLC/PRF/5 poly A^+ RNAs containing HBV sequences (17S and 19S) might be located in the $Hinc_3$ fragment region of the HBV genome. In order to demonstrate that the RNA initiation site for transcription of these poly A^+ RNAs (17S and 19S) resides in this fragment, we developed a "5'-poly A^+ RNA protection assay" on Southern blots. Different HBV DNA fragments were blotted to DBM paper and hybridized to [^{32}P] 5'-end-labelled cytoplasmic poly A^+ RNA.

Figure 4A shows ethidium bromide stained viral DNA fragments prior to transfer to DBM paper. Lane A represents the 1.35 kb Bam HI HBV DNA fragment digested with Hpa I (producing 0.95 kb and 0.40 kb fragments (see Figure 1)). Lane B shows the two fragments (1.5 kb and 0.35 kb) produced after digestion of the 1.85 kb Bam HI fragment with Taq I which, under the conditions used, cleaved only the restriction site located nearer to the EcoRI site (see Figure 1A). The blotted fragments were subsequently hybridized to [^{32}P] 5'-end-labelled cytoplasmic poly A^+ RNA from PLC/PRF/5 cells. After hybridization, the DBM paper was treated with pancreatic RNAse, so that only the viral DNA fragment containing the RNA initiation site would protect the [^{32}P] 5'-end of the RNA and generate a hybridization signal. As shown in Figure 4B, only the 1.5 kb Taq I fragment protected the 5'-end-labelled RNA. In conjunction with the mapping studies (Figures 2 and 3), this provides additional evidence that the *in vivo* RNA initiation site for transcription of both 17S and 19S poly A^+ RNAs containing viral sequences is located between coordinates 2700 bp and 2900 bp on the HBV genome (see Figure 1).

From the above experiments, a schematic diagram mapping the origin and orientation of the 17S and 19S RNAs containing HBV sequences can be constructed (Figure 5). The bulk of these RNAs consist of the same HBV DNA sequences ($\sim$ 1750 nucleotides), but there are 600 nucleotides in the 19S poly A^+ RNA which cannot be accounted for by the mapping studies. These extra sequences are located at the 3' end of the 19S RNA molecule, suggesting a "read-through" transcription into cellular sequences. Since the 5' ends of the other 3 poly A^+ viral RNAs of lower abundance (containing HBcAg gene sequences) were not located on the HBV map, these RNAs have not been included on this diagram.

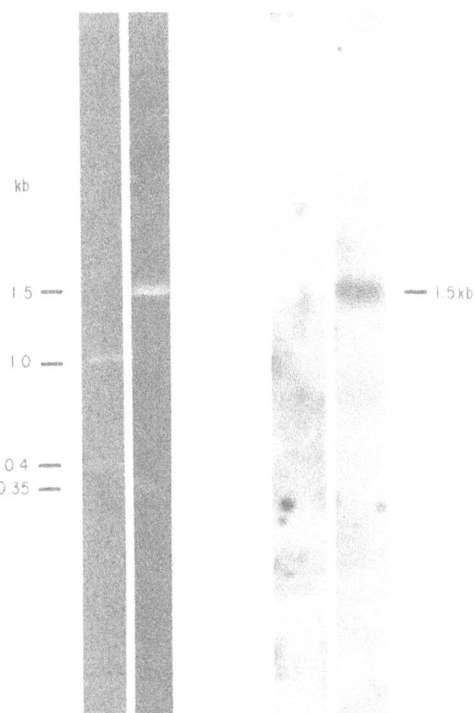

Figure 4. Identification of putative in vivo RNA initiation site(s) for the 17S and 19S poly A^+ RNAs of the PLC/PRF/5 cell line using a 5'-[^{32}P]-labelled poly A^+ RNA protection assay. Panel A shows the EtBr stained HBV DNA fragments after digestion with specific restriction enzymes and separation on a 0.8% agarose slab gel. Lane A, 1 μg of the 1.35 kb Bam H1 fragment digested with Hpa I; lane B, 1 μg of the 1.85 kb Bam H1 fragment digested with Taq I. The DNA fragments in lanes A and B were subsequently transferred to DBM-cellulose paper and hybridized with 5'-[^{32}P]-end-labelled poly A^+ RNA. 5'-[^{32}P]-labelling of poly A^+ RNA was performed as described in MATERIALS AND METHODS. Hybridization was performed under standard conditions with 3 x 10^7 cpm of TCA precipitable radioactivity (specific activity 1-2 x 10^6 cpm/μg). After hybridization, the paper was washed and treated with pancreatic RNAse as described in MATERIALS AND METHODS. The autoradiogram was developed after 10 days of exposure as shown in Panel B.

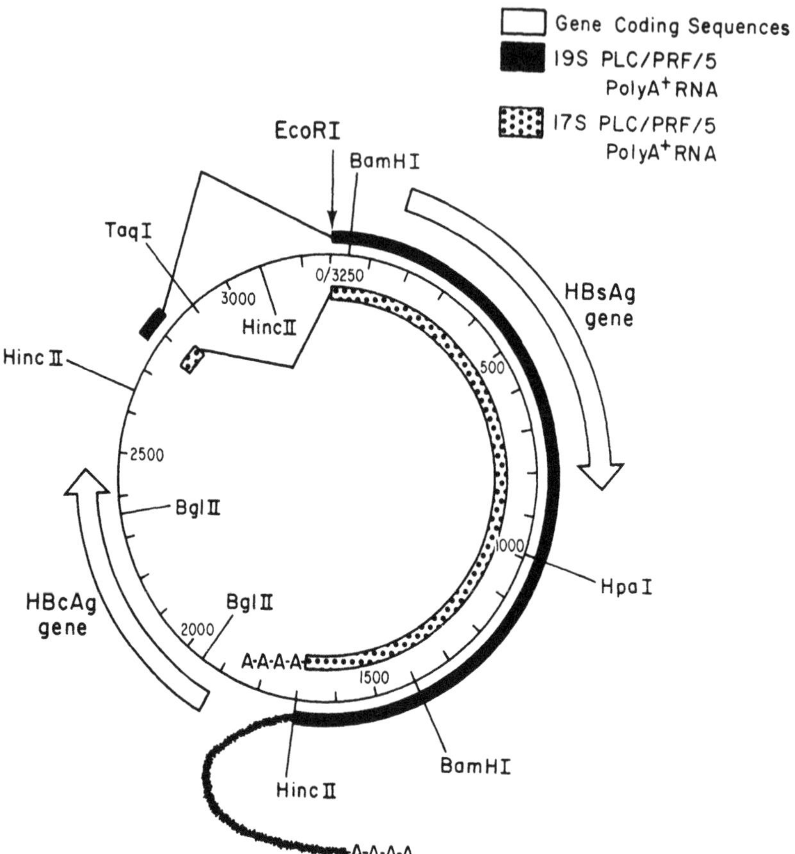

Figure 5. Schematic diagram illustrating the genetic organization and expression of HBV DNA sequences for 17S and 19S poly A^+ RNAs in the PLC/PRF/5 cell line. Restriction endonuclease sites that were relevant to this study are shown. The two small boxes between the map coordinates 2700 and 2900 bp are the proposed leader sequence for the 17S and 19S poly A^+ RNAs. The lines joining the leader sequence to the main body of the RNAs represent the presumptive splicing events. The "saw tooth" line at the 3' end of 19S poly A^+ RNA indicates the proposed non-viral sequence linked to viral RNA.

Expression of viral RNAs in the liver of chimpanzee HBV carriers

Parallel studies have been performed with RNA isolated from the liver of chimpanzee chronic carriers of hepatitis B virus. To identify RNA molecules derived from viral genes in infected chimpanzee liver, RNA blots were hybridized with specific HBV DNA fragment probes as illustrated in Figure 1. Although a diffuse hybridization pattern was observed (Figure 6), three predominant poly A^+ RNAs of 24S, 19S and 14S were detected with a full-length HBV DNA probe (lane A). Comparable amounts of poly A^+ RNA from uninfected chimpanzee liver showed no hybridization (lanes B and C), indicating specificity of the probe. The 24S poly A^+ RNA ($\sim$ 3400 nucleotides) hybridized to both HBsAg and HBcAg gene

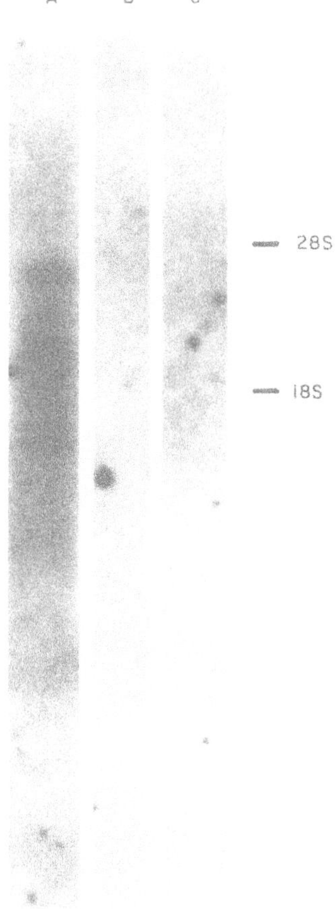

Figure 6. Identification of HBV DNA sequences in poly A^+ RNA isolated from the liver of a chimpanzee HBV carrier. Isolation of poly A^+ RNA from the liver of a chimpanzee HBV carrier (lane A) and non-carriers (lanes B and C), glyoxal denaturation of the poly A^+ RNA, electrophoresis, blotting of the RNA to DBM paper and hybridization was performed as mentioned in MATERIALS AND METHODS. Two ug of poly A^+ RNA was applied to each lane. [^{32}P]-labelled 3250 bp HBV DNA at a specific activity of 2 x 10^8 cpm per ug DNA was utilized as hybridization probe.

sequence probes (Figure 7, lanes A and B) as well as all other subgenomic probes of Figure 1B (data not shown). Chimpanzee 19S poly A^+ RNA ($\sim$2400 nucleotides) also hybridized to both HBsAg and HBcAg gene sequence probes (Figure 7, lanes A and B). Due to partial degradation of RNA in this sample, specific characterization of the 14S RNA containing HBV sequences was not possible.

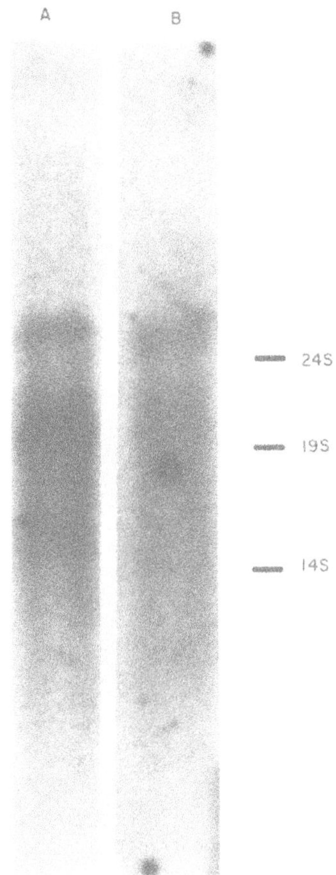

Figure 7. Identification of specific HBV DNA sequences in poly A^+ RNA isolated from the liver of a chimpanzee HBV carrier using HBsAg and HBcAg gene fragment probes. Isolation of poly A^+ RNA from the liver of a chimpanzee HBV carrier, glyoxal denaturation of the poly A^+ RNA, electrophoresis, blotting of the RNA to DBM paper and hybridization was performed as mentioned in MATERIALS AND METHODS. Two ug of poly A^+ RNA was applied to each lane. Separately, lane A was hybridized with a [^{32}P]-labelled HBsAg gene sequence probe and lane B with a [^{32}P]-labelled HBcAg gene sequence probe.

DISCUSSION

Natural or experimental transmission of hepatitis B virus (HBV) has been achieved only in humans and chimpanzees, and numerous attempts to propagate HBV in tissue culture have been unsuccessful. These factors have impeded understanding the biology of hepatitis B virus, and details concerning the mechanism of viral morphogenesis have remained obscure. In separate studies, chimpanzee HBV carriers have been examined and uniquely integrated HBV DNA molecules have not been detected (18,19,20). In contrast, it has been shown that PLC/PRF/5 cells, which produce only one viral protein, HBsAg, contain as many as six HBV genomes integrated into host cell DNA (1,2,3,6). In PLC/PRF/5 cells, no HBV DNA sequences were detected in extrachromosomal form. Therefore, it was relevant to examine RNA expression of the HBV genome in chimpanzee liver in comparison to the PLC/PRF/5 cell line, because specific HBV transcripts might represent unique RNA products involved in various aspects of viral infection and/or tumor development.

The present study suggests that the nature of HBV viral transcription depends on the integration state of viral DNA. In the chimpanzee model in which HBV DNA is not integrated and all the known viral products are made, three specific poly A^+ HBV RNA transcripts were detected. These transcripts (24S, 19S and 14S) are derived from a free viral genome, and the larger RNA appears to contain the full complement of the HBV genome. The PLC/PRF/5 cell line contains 5 poly A^+ viral RNA transcript (37S, 26S, 21S, 19S and 17S), and we suspect that all but the 17S viral RNA may represent viral-cellular cotranscripts. The 19S poly A^+ viral RNA transcript in PLC/PRF/5 cells is clearly distinct from the 19S poly A^+ viral RNA of chimpanzee HBV carrier liver. The latter contains both HBsAg and HBcAg gene coding sequences, whereas the former contains only HBsAg coding sequences. Other small specific viral RNA transcripts, one of which migrates at 14S on a denaturing agarose gel, are present in chimpanzee HBV carrier liver, but for technical reasons their characterization has not been satisfactory.

Recently, Summers and Mason (21) reported a full-length viral poly A^+ RNA in the duck hepatitis B virus (DHBV) system and have proposed that this transcript serves as an intermediate for asymmetric replication of DHBV via a reverse transcriptase-like reaction. Their proposed mechanism requires initial synthesis of a full-length viral (+) strand RNA transcript which they have identified ("pregenome" RNA) and which serves as a template for synthesis of the virus DNA (-) strand. Chimpanzee liver 24S poly A^+ viral RNA may represent another example of a viral "pregenome" in the hepatitis B or "Hepadna" virus family (22). It is also interesting that both the chimpanzee and duck infected liver systems contain covalently-closed supercoiled HBV genomes which could serve as a template for synthesis of a full-length genome transcript (19,21) and that specific integrated HBV DNA molecules have not been identified in either system. However, in both systems, other roles for these full-length poly A^+ RNA transcripts in viral mRNA biogenesis (processing) have not been excluded.

RNA mapping studies suggest that a viral RNA initiation site (promoter) located at map position ∿ 2800 bp is utilized for transcription of PLC/PRF/5 17S and 19S poly A^+ RNAs. Such results are consistent with the finding of a Hogness-like sequence in the HBV DNA genome at position 2776 (8,23) and evidence from mammalian cell transfection studies, using recombinant plasmids containing portions of the HBV genome, that the HBsAg gene promoter is located upstream to the EcoRI site (8). Therefore, we conclude that this RNA initiation site is the promoter utilized *in vivo* for HBsAg gene expression in PLC/PRF/5 cells. However, we have not formally ruled out the possibility that HBsAg RNA initiation begins within the $Hinc_1$ fragment (map coordinates 1.4-1.7 kb), as would occur if HBsAg RNA transcription were both initiated and terminated within this region and multiple splicing events occurred.

The present study suggests that a splicing event might be involved in the maturation of 17S and 19S poly A^+ RNAs in PLC/PRF/5 cells. The RNA mapping experiments indicate that a region between coordinate positions 2900 bp and 3250/0 bp is not present in these cytoplasmic viral RNAs. This finding excludes the possibility that the HBsAg polypeptide is cleaved from a precursor protein corresponding to the entire region S as suggested previously (17). Either splicing of a primary gene transcript or deletion of this portion of the HBV genome could explain this finding, although the latter is less likely in view of the absence of this sequence in both 17S and 19S poly A^+ RNAs. Although specific data was not presented, Gough and Murray (24) recently postulated that processing of high molecular weight HBV DNA transcripts is required for biogenesis of HBcAg and HBeAg mRNAs. Three high molecular weight RNA transcripts identified in the present study, one in PLC/PRF/5 cells (37S) and two in chimpanzee (24S and 19S), could serve as potential mRNA precursors, but we have not so far identified a role for these transcripts in specific HBV mRNA biogenesis.

Quantitative estimation of the hybridized subgenomic HBV fragments indicates that the major HBsAg sequence containing PLC/PRF/5 poly A^+ RNAs (1800 and 2400 nucleotides) share ∿ 1750 nucleotides of viral sequences. The additional ∿ 600 nucleotides present in the 19S poly A^+ RNA (2400 nucleotides) appear not to represent viral sequences. This should be considered a working hypothesis, since these extra sequences might represent a partial duplication of the HBV genome within this region. It is most likely, however, that these additional sequences (located at the 3'-end of 19S poly A^+ RNA) are derived by cotranscription of viral-cellular sequences via a "read through" mechanism (see Figure 5). This suggests further that these two poly A^+ RNAs may be transcribed from two distinct integrated HBV DNA molecules.

Edman et al. (3) have suggested that integration of HBV in PLC/PRF/5 cells in most cases occurs at the "nicked" cohesive end region of the virus located 1400-1700 bp clockwise from the EcoRI site (17,23). They also suggested that one of the integrated sequences might represent a tandem integration of the virus, which has also been found in primary liver cancers (6). Thus, it is possible that the 17S poly A^+ RNA (which contains only viral sequences) is transcribed from a "tandem-integrated" transcription unit distinct from the transcriptional unit which generates the 19S poly A^+ RNA putative cotranscript. It should be noted that the 3' proximal subgenomic HBV DNA fragment probe ($Hinc_1$, Figure 3), which hybridized to the 19S poly A^+ RNA, terminates at, or very close to, the position of the "nicked" cohesive end region of the virus. Nevertheless, other possible mechanisms for the biogenesis of both 17S and 19S poly A^+ RNAs cannot be excluded, since the arrangement of integrated HBV DNA sequences within host chromosomal DNA in PLC/PRF/5 cells has not yet been elucidated.

In retrovirus systems, recent reports show that cotranscription of viral and cellular sequences occurs during oncogenic transformation by a "promoter insertion mechanism" (25,26), and evidence for high molecular weight viral-cellular RNA cotranscripts from integrated viral DNA has been reported (25,26). During the present study, we detected two high molecular weight RNAs ($\sim$ 21S and 26S) containing only a small region of the HBV genome within the core gene segment. Their size suggests that they might represent viral-cellular cotranscripts. Although their abundance was very low as compared to 17S and 19S poly A^+ RNAs, their presence in PLC/PRF/5 might be relevant to the transformed state of this cell line. Since both of these RNAs contain core antigen gene sequences and we have identified a potential promoter for the core gene in the "cohesive end" region of the HBV genome by _in vitro_ transcription (27, our unpublished observations), it is possible that this promoter is utilized for synthesis of these RNAs. This system, therefore, may have properties similar to retrovirus transcription where direct repeat sequences have been identified in the promoter region of specific oncogenes; for review, see Varmus (28).

Although a total of 5 poly A^+ RNA transcripts containing HBV sequences were identified, in the present study our efforts were concentrated on characterizing the 17S and 19S RNAs because

1) they were the most abundant and 2) the PLC/PRF/5 cell line produces only one identifiable viral gene product, HBsAg, for which the coding sequences are present in these RNAs. The less abundant RNA transcripts containing HBcAg gene sequences will be more amenable to study when other viral polypeptides related to viral morphogenesis or cellular transformation become identified. Although the core gene sequence in these RNAs does not appear to be related to core antigen expression in PLC/PRF/5 cells (HBcAg is not produced), sequencing studies (23) suggest that short direct repeats and a potential promoter signal is located in the "cohesive-end" region just proximal to the core gene. Thus, it will be important to elucidate the structural nature and coding properties of these less abundant high molecular weight poly A^+ RNAs containing HBcAg gene sequences.

ACKNOWLEDGMENTS

We thank Dr. Jesse Summers, Institute for Cancer Research, Fox Chase Cancer Center, Philadelphia, PA, for originally supplying the recombinant plasmid $_p$A01-HBV and Dr. P. Sarkar for advice in [^{32}P] RNA 5'-end labelling studies. This research was supported in part by National Institutes of Health grants AM 17609 and CA 32605, the Sarah Chait Memorial Foundation and the Gail I. Zuckerman Foundation. These studies represent part of a thesis by N.R.O. submitted to the Sue Golding Graduate Division, Albert Einstein College of Medicine, in partial fulfillment of the requirement for the Ph.D. degree. Thanks to Ms. Michelle O'Donnell for typing of this manuscript.

REFERENCES

1. Marion, P. L., Salazar, F. H., Alexander, J. J., and Robinson, W. S. (1980). State of hepatitis B viral DNA in a human hepatoma cell line. J. Virol. 33: 795-806.
2. Chakraborty, P. R., Ruiz-Opazo, N., Shouval, D., and Shafritz, D. A. (1980). Identification of integrated hepatitis B virus DNA and expression of viral RNA in a HBsAg producing human hepatocellular carcinoma cell line. Nature 286:531-533.
3. Edman, J. C., Gray, P., Valenzuela, P., Rall, L. B., and Rutter, W. J. (1980). Integration of hepatitis B virus sequences and their expression in a hepatocellular carcinoma. Nature 286:535-537.
4. Aden, D. P., Fogel, A., Plotkin, S., Damjanov, I., and Knowles, B. B. (1979). Controlled synthesis of HBsAg in a differentiated human liver carcinoma-derived cell line. Nature 282:615-616.

5. Dubois, M. F., Pourcel, C., Rousset, S., Chany, C., and Tiollais, P. (1980). Excretion of hepatitis B surface antigen particles from mouse cells transformed with cloned viral DNA. Proc. Natl. Acad. Sci. USA 77:4549-4553.
6. Brechot, C., Pourcel, C., Louise, A., Rain, B., and Tiollais, P. (1980). Presence of integrated hepatitis B virus DNA sequences in cellular DNA of human hepatocellular carcinoma. Nature 286:533-535.
7. Twist, E. M., Clark, H. F., Aden, D. P., Knowles, B. B., and Plotkin, S. A. (1981). Integration pattern of hepatitis B virus DNA sequences in human hepatoma cell lines. J. Virol. 37:239-243.
8. Pourcel, C., Louise, A., Gervais, M., Chenciner, N., Dubois, M. F., and Tiollais, P. (1982). Transcription of the hepatitis B surface antigen gene in mouse cells transformed with cloned viral DNA. J. Virol. 42: 100-105.
9. Shouval, D., Reid, L. M., Chakraborty, P. R., Ruiz-Opazo, N., Morecki, R., Gerber, M. A., Thung, S. N., and Shafritz, D. A. (1981). Tumorigenicity in nude mice of a human hepatoma cell line containing hepatitis B virus DNA. Cancer Res. 41:1342-1350.
10. Innis, M. A. and Miller, D. L. (1979). Alpha-fetoprotein gene expression, control of alpha-fetoprotein mRNA levels in cultured rat hepatoma cells. J. Biol. Chem. 254:9148-9154.
11. Aviv, H. and Leder, P. (1972). Purification of biologically active globin messenger RNA by chromatography on oligothymidylic acid-cellulose. Proc. Natl. Acad. Sci. USA 69:1408-1412.
12. McMaster, G. K. and Carmichael, G. G. (1977). Analysis of single- and double-stranded nucleic acids on polyacrylamide and agarose gels by using glyoxal and acridine orange. Proc. Natl. Acad. Sci. USA 74: 4835-4838.
13. Dingman, C. W. and Peacock, A. C. (1968). Analytical studies on nuclear ribonucleic acid using polyacrylamide gel electrophoresis. Biochemistry 7:659-667.
14. Alwine, J. C., Kemp, D. J., and Stark, G. R. (1977). Method for detection of specific RNA in agarose gels by transfer to diazobenzyloxy-methyl-paper and hybridization with DNA probes. Proc. Natl. Acad. Sci. USA 74: 5350-5354.
15. Shatkin, A. J. (1976). Capping of eukaryotic mRNAs. Cell 9: 645-653.
16. Shimski, H., Miwa, M., Kato, K., Noguchi, M., Matasushima, T., and Sugimura, T. (1976). A novel phosphodiesterase from cultured tobacco cells. Biochemistry 15:2185-2190.

17. Tiollais, P., Charnay, P., and Vyas, G. N. (1981). Biology of hepatitis B virus. Science 213:406-411.
18. Shouval, D., Chakraborty, P. R., Ruiz-Opazo, N., Baum, S., Spigland, I., Muchmore, E., Gerber, M. A., Thung, S. N., Popper, H., and Shafritz, D. A. (1980). Chronic hepatitis in chimpanzee carriers of hepatitis B virus: Morphologic, immunologic, and viral DNA studies. Proc. Natl. Acad. Sci. USA 77:6147-6151.
19. Ruiz-Opazo, N., Chakraborty, P. R., and Shafritz, D. A. (1982). Evidence for supercoiled hepatitis B virus DNA in chimpanzee liver and serum Dane particles: Possible implications in persistent HBV infection. Cell 29:129-138.
20. Monjardino, J., Fowler, M. J. F., Montano, L., Weller, I., Tsiquaye, K. N., Zuckerman, A. J., Jones, D. M., and Thomas, H. C. (1982). Analysis of hepatitis virus DNA in the liver and serum of HBe antigen positive chimpanzee carriers. J. Med. Virol. 9:189-199.
21. Summers, J. and Mason, W. S. (1982). Replication of the genome of a hepatitis B-like virus by reverse transcription of an RNA intermediate. Cell 29:403-415.
22. Robinson, W. S., Marion, P., Feitelson, M., and Siddiqui, A. (1982). *in* Viral Hepatitis, 1981 International Symposium, W. Szmuness, H. J. Alter and J. E. Maynard, eds., Franklin Institute Press, Philadelphia, pp. 57-68.
23. Valenzuela, P., Quiroga, M., Zaldivar, J., Gray, P., and Rutter, W. J. (1980). The nucleotide sequence of the hepatitis B viral genome and the identification of the major viral genes. *in* Animal Virus Genetics, B. Fields, R. Jacnisch and C. F. Fox, eds., Academic Press, New York, pp. 57-70.
24. Gough, N. M. and Murray, K. (1982). Expression of the hepatitis B virus surface core and e antigen genes by stable rat and mouse cell lines. J. Molec. Biol. 162:43-67.
25. Neel, B. G., Hayward, W. S., Robinson, H. L., Fang, J., and Astrin, S. M. (1981). Avian leukosis virus-induced tumors have common proviral integration sites and synthesize discrete new RNAs: Oncogenesis by promoter insertion. Cell. 23:323-334.
26. Hayward, W. S., Neel, B. G., and Astrin, S. M. (1981). Activation of a cellular *onc* gene by promoter insertion in ALV-induced lymphoid leukosis. Nature 290:475-480.
27. Chakraborty, P. R., Ruiz-Opazo, N., and Shafritz, D. A. (1981). Transcription of human hepatitis B virus core antigen gene sequences in an *in vitro* cellular extract. Virology 111:647-652.
28. Varmus, H. E. (1982). Form and functions of retroviral proviruses. Science 216:812-820.

HEPATITIS B SURFACE ANTIGEN POLYPEPTIDE AND SYNTHETIC PEPTIDE VACCINES

G. R. Dreesman, I. Ionescu-Matiu, Y. Sanchez,
R. C. Kennedy, J. T. Sparrow and J. L. Melnick

Department of Virology and Epidemiology
Baylor College of Medicine
Houston, Texas 77030

The hepatitis B vaccine recently licensed for use in the United States has undergone several field trials in high-risk populations and has proven to be immunogenic, efficient and safe. The hepatitis vaccine is unique in that the hepatitis B virus (HBV), the ultimate source of the vaccine, cannot be grown in tissue culture. The immunizing antigen, purified from plasma of healthy human carriers of HBV, consists of noninfectious 22 nm particles of hepatitis B surface antigen (HBsAg), produced as excess viral surface protein and released in the bloodstream. Although the HBsAg particles are highly purified and inactivated, there is some concern that the vaccine may contain traces of host material or infective adventitious agents, which may trigger undesirable effects in the recipients. Ideally, a hepatitis B vaccine candidate should have the following properties: 1) be free of normal human serum protein contaminants, 2) be free of potential residual infectious HBV, 3) be free of other viruses and nucleic acid, 4) have a high degree of immunogenicity when administered in conjunction with an adjuvant suitable for use in humans, 5) be independent of human HBV carriers as a source of immunogenic material, 6) have a reproducible composition, 7) cost less than the present vaccine, and 8) be available in unlimited supplies.

To comply with part of the above, we proposed and tested a potential "second generation" vaccine, composed only of viral specific proteins. We compared the immunogenicity of three HBsAg preparations, all derived from plasma of human HBV carriers. The first preparation consisted of intact 22 nm HBsAg particles, purified by sedimentation in the ultracentrifuge followed by isopycnic and zonal banding in CsCl (1). The second preparation consisted of a pool of viral specific proteins derived by preparative polyacrylamide gel electrophoresis from purified HBsAg particles disrupted with sodium dodecyl sulfate (SDS) and 2-mercaptoethanol (2). The polypeptide profile of different subtypes of HBsAg--adw, ayw, adr--was similar and remarkably reproducible from one preparation to the next. They contained two major components, with molecular weights of 25,000 and 30,000, generally referred to in the literature as P25 and GP30, respectively. GP30 is the glycosylated form of P25. Together, these components represent over 65% of the mass of HBsAg. They are viral specific proteins in that they do not cross-react with any human components (2). Cross-reactivity, particularly with albumin and IgG, was associated with the higher molecular weight polypeptides (3).

An SDS-denatured P25 + GP30 pool, prepared under reducing conditions, was shown to be immunogenic in chimpanzees and to confer protection to challenge with infectious HBV (4). One of the major deficiencies of this type of preparation was its low immunogenicity. Early work in our laboratory (5), as well as others (see reference 6 for review), demonstrated that the antigenic determinants associated with HBsAg are conformation dependent. Reduction of disulfide bonds and alkylation of free thiol groups destroyed virtually all the immunogenic and antigenic activity of HBsAg particles. Our P22 + GP30 polypeptide pool was prepared under reducing conditions, and this resulted in a low immunogenicity. However, it should be pointed out that the polypeptide pool was not alkylated. Therefore, reoxidation and reassociation of some of the free thiol groups may have resulted in restoration of part of the native conformation, which in turn conferred low levels of immunogenicity to this preparation.

The third vaccine candidate consisted of a mixture of P25 + GP30 obtained under nonreducing conditions, and thus in a non-denatured form (7). We used a protocol first developed for Semliki Forest virus by Simons et al. (8) and later applied to HBsAg by Zuckerman et al. (9). Purified HBsAg particles were disrupted with a nonionic detergent (Triton X-100). In the presence of detergent, polymers of P25 + GP30 were separated by affinity chromatography on lentil lectin. Finally, removal of detergent by isopycnic banding in CsCl resulted in a product which contained aggregated protein complexes referred to as micelles. Micelles of P25 + GP30 ranged in diameter from 50 to 150 nm and had a buoyant density in sucrose of 1.19 g/cm^3.

Groups of 6 BALB/c mice were immunized with each of the three preparations (10 μg of viral protein per dose), administered either as saline suspension or adsorbed on aluminum gel (7). The antibody response was monitored by a micro solid phase radioimmunoassay (micro-SPRIA) (10).

No response was obtained after the primary inoculation of the aqueous suspension of SDS-denatured polypeptides and only two out of six mice responded with low titers (1:10) after the booster. All mice produced detectable antibody levels after a primary inoculation of saline-suspended intact HBsAg particles or micelles. In both groups, the antibody titers increased approximately 25-fold after booster. The titers induced with micelles were significantly higher than those induced with intact HBsAg particles, both after the primary and after the booster inoculation.

The immunogenicity of all three preparations was enhanced when inoculated as an alum-precipitate (Table 1). All six mice responded to the SDS-denatured preparation, but the relative degree of immunogenicity was significantly lower than that of the other two vaccines. The micelles were again the most immunogenic preparation, but this difference was noted only after the booster inoculation, as evidenced by the fact that the geometric mean titers increased approximately 160-fold in mice immunized with micelles and only 3-fold in those injected with intact HBsAg (Table 1). Two-hundred

Table 1. Mean anti-HBs titers in mice[a] inoculated with alum-adsorbed HBsAg vaccine preparations

HBsAg Preparation	Days after primary inoculation		Days after booster inoculation	
	14	25	14	35
Intact 22 nm HBsAg particles	3655[b]	6250	53,000	6250
SDS-denatured P25 + GP30 pool	< 10	< 10	85	17
Nondenatured P25 + GP30 micelles	6250	8137	1,300,000	270,000
SP1-tetanus toxoid conjugate	ND[c]	ND	6250	5500

[a]Groups of six mice were inoculated with each designated preparation on days 0 and 35.

[b]Reciprocal of the arithmetic mean antibody titer of the six mice in each group, determined by micro-SPRIA. Five-fold antisera dilutions were added to HBsAg/adw coated wells. Affinity purified ^{125}I-goat IgG anti-mouse IgG was used as detecting probe.

[c]ND = not done.

days after the booster, the anti-HBs titers were still high in mice immunized with either aqueous or alum-adsorbed micelles, whereas the antibodies were no longer detectable in mice immunized with SDS-denatured polypeptides adsorbed on alum gel. The mice immunized with HBsAg particles were no longer available at this time period.

The above data demonstrated that the micelles have a higher immunogenicity than the intact HBsAg particles. Also, the antibody response was characterized by a slow drop over an extended period of 200 days. However, several disadvantages are inherently associated with this type of preparation: it remains dependent on plasma obtained from HBV carriers for source of viral material, approximately 40% of the starting purified HBsAg protein is lost during the preparation process, and the additional purification procedures can only add to the already high cost of the present vaccine.

Recently, we investigated another approach toward the development of an HBV vaccine candidate, namely a synthetic vaccine. The background is as follows. DNA extracted from hepatitis B Dane particles had been cloned in bacteria using plasmid vectors. A sequence of 892 base pairs, designated S gene, had been found to encode the major polypeptide (P25) associated with HBsAg. The amino acid sequence of P25 had been deduced from the DNA sequence (11,12, 13) and partially confirmed by amino acid analysis of purified P25 (14). Computer analysis of the amino acid sequence predicted hydrophilic and hydrophobic regions of the molecule (15) as well as the secondary structure (16). The hydrophilic regions (amino acid residues 43 through 65 and 109 through 145) were postulated to be exposed on the surface of the intact particle. In addition, β-turns in these areas suggested potentially exposed antigenic determinants. During the past year, several researchers (17,18,19) synthesized small, linear, non-overlapping analogues of the hydrophilic regions of P25. Several of these peptides proved to be immunogenic in experimental animals when coupled to a protein or red blood cell carrier.

We selected for synthesis region 117-137 of the amino acid sequence reported by Charnay for P25 subtype *ayw* (11). This sequence was located in a hydrophilic area and a cyclic peptide could be formed by introduction of a disulfide bond between cysteine residues 124 and 137 (20). In this way, we expected to obtain a peptide more "native" in conformation and therefore having a putative high degree of immunogenic activity.

Two cyclic disulfides were synthesized. The first (referred to as SP1) contained amino acid residues 122 through 137 and the second (referred to as SP2) contained amino acid residues 117 through 137. Both synthetic peptides were incorporated into different adjuvants (Freund's complete adjuvant, aluminum gel and liposomes) without being linked to any protein carrier, and inoculated (50 μg peptide per dose) into groups of 6 BALB/c mice (20). The animals were bled 21 days after the primary inoculation, boosted on day 22 and bled 10 days after the booster inoculation. Both synthetic peptides induced anti-HBs antibodies in mice. Liposomes were the most efficient adjuvant vehicle, aluminum gel the poorest. Fifty percent of animals in the groups inoculated with the peptide-liposome preparations responded after one inoculation, and the titers increased significantly after the booster. The unexpectedly high immunogenicity of the cyclic peptides may be due to the presence of a major immunogenic determinant, or to the locking of the secondary structure of a potent immunogen by the disulfide cyclization.

We tried to enhance the immunogenicity of SP1 by procedures that would be acceptable for future field trials in humans (21). Peptide micelles were prepared using a procedure similar to that described previously for virus proteins. These aggregates were smaller (40 to 80 nm diameter) and had a lower density (1.10 g/cm^3 in sucrose) than micelles prepared from solubilized HBsAg particles. In addition, a conjugate was obtained by covalent coupling of SP1 to a protein carrier-tetanus toxoid-using carbodiimide as crosslinker. Both the synthetic peptide micelles and the synthetic peptide-tetanus toxoid conjugate (SP1-TT) were tested for immunogenicity in mice, either as an aqueous suspension or as alum gel precipitate. The relative immunogenicity of the four preparations, illustrated by the number of mice developing an anti-HBs response and by the anti-HBs levels, is summarized in Table 2. After the primary inoculation, the response appeared to be no better than after liposome-entrapped, uncoupled synthetic peptides. However, after a booster inoculation, 80-100% of the animals responded to synthetic peptide micelles precipitated on alum gel and to both formulations of SP1-TT conjugate. The mice immunized with the SP1-TT responded with high titers, the peptide micelles precipitated on alum gel induced relatively low levels of anti-HBs, while the peptide micelles in saline were a weak immunogen.

The arithmetic mean anti-HBs titers in mice immunized with the SP1-TT conjugate precipitated on alum gel was compared to the titers induced by intact HBsAg particles, SDS-denatured P25 + GP30 polypeptides and P25 + GP30 micelles (Table 1). The P25 + GP30 micelle preparation derived under non-denaturing conditions from HBsAg particles was clearly the superior immunogen. It was noteworthy that after one booster inoculation, the arithmetic mean titer obtained with the SP1-TT conjugate was comparable to that induced with intact HBsAg particles and showed only a small decrease in antibody activity on day 35 (Table 1). We are presently optimizing our SP1-TT conjugate by testing different

Table 2. Immunogenicity in mice of SP1 in micelle form or conjugated to tetanus toxoid

	SP1 micelles, saline	SP1 micelles, alum	SP1-TT, saline	SP1-TT, alum
32 days after primary inoculation	0/6[a]	2/6	3/6	1/6
32 days after booster inoculation	1/6 (2.7)[b]	4/5 (7.5-76)	4/5 (3.4-153)	6/6 (2.1-123)

[a]Number of mice responding/number of mice inoculated. All sera were tested at a 1:6 dilution by the AUSAB test (Abbott Laboratories, North Chicago, IL). Sera with S/N values $\leq$ 2.1 were considered negative.

[b]Numbers in parentheses represent the range of S/N in the responding animals.

crosslinkers, as well as different SP1-TT conjugation ratios.

A major aspect in selecting a synthetic peptide to use as a potential HBV vaccine is to ascertain the specificity of the antigenic determinants, or epitopes, present on that peptide. It is known that the cross-reacting group antigenic determinant a of HBsAg is of prime importance in conferring immunity, since anti-a antibodies induced by immunization with one serotype confer protection against reinfection with the other HBV serotypes. Therefore, a synthetic peptide suitable as an HBV vaccine candidate should contain the a epitope. To determine which HBsAg antigenic determinants are present on SP1, we assessed the ability of the peptide to react with a panel of anti-HBs monoclonal antibodies of known specificity, produced and characterized in our laboratory (22). Cyclic SP1 reacted with 5 of 13 anti-a monoclonal antibodies and failed to react with the remaining 8 (Table 3) (23). This indicated that SP1 contains an a epitope and also that the a specificity associated with HBsAg has one or more antigenic determinants, unrelated to that expressed on SP1. After reduction and alkylation of cyclic SP1, the resulting linear peptide no longer reacted with the 5 anti-a monoclonal antibodies. This demonstrated that the disulfide bond is critical in conferring the a specificity and that the a epitope present on cyclic SP1 is

Table 3. Analysis of SP1 for HBsAg group and subtype antigenic determinants utilizing a panel of monoclonal antibodies

Monoclonal antibody	Reactivity of the monoclonal antibody with: HBsAg[a]			SP1[b]	
	adw	ayw	adr	cyclic	linear
anti-a (5)[c]	+	+	+	+	-
anti-a (8)	+	+	+	-	-
anti-y (3)	-	+	±[d]	+	+
anti-w (1)	+	+	-	-	-

[a]Determined by micro-SPRIA. Wells of a polyvinyl microtiter plate were coated with purified 22 nm HBsAg particles of the designated subtype; monoclonal antibody was then added; affinity purified ^{125}I-goat IgG anti-mouse IgG was used as detecting probe.

[b]Determined by the ability of SP1 to inhibit the reaction of the monoclonal antibody with the HBsAg on the solid phase. An optimal predetermined dilution of each monoclonal antibody was first incubated with 80 μg of either cyclic or linear SP1. The mixture was then tested for residual anti-HBs activity by micro-SPRIA, as described in "a."

[c]Numbers in parentheses represent the number of individual monoclonal antibodies tested.

[d]Two of the three monoclonal antibodies characterized as having anti-y specificity also reacted with the adr subtype. Preincubation of these antibodies with cyclic or linear peptide inhibited the reaction with HBsAg subtype ayw, but did not affect the reactivity with adr subtype.

conformation-dependent. SP1 also reacted with three of three anti-y monoclonal antibodies (Table 3). Reduction of the intrachain disulfide bond and alkylation of free thiol groups did not affect this reactivity, indicating that SP1 contains a sequential y epitope. SP1 did not react with one monoclonal antibody with anti-w specificity. The presence of a y epitope on SP1 was not unexpected. Two differences in the amino acid residues of P25 subtype ayw compared to subtype adw were reported by Peterson in the 122-127 region (24). The ayw contains threonine and tyrosine, while adw contains asparagine and phenylalanine in positions 131 and 134, respectively. These amino acid substitutions were thought to reflect the subtype y and d activities. This was confirmed by our detection of an epitope with y specificity on SP1 (amino acid residues 122-137).

The next question we posed was: Does SP1 react with antibodies produced by humans in response to a natural HBV infection? To answer this question, we analyzed our synthetic peptide for its ability to inhibit a HBsAg specific idiotype-anti-idiotype reaction. The idiotype of an antibody is associated with the variable region of the immunoglobulin molecule (i.e. human anti-HBs) which contains the antigen-combining site. We prepared an anti-idiotype reagent by immunizing rabbits with specifically purified human anti-HBs IgG (25). The anti-idiotype antiserum was extensively adsorbed with normal human serum, so that the final preparation recognized only the antigen-combining region of the anti-HBs IgG molecule (the idiotype). Since the anti-idiotype antiserum is specifically directed against the antibody combining site of the idiotype, preincubation of the idiotype with HBsAg or SP1 should inhibit the reaction between the idiotype and the anti-idiotype antibody. Our initial studies indicated that humans infected with HBV produce anti-HBs which contain a common human idiotype (CHId) because the configuration of human anti-HBs idiotype was similar in different individuals (25).

The CHId was found to be associated with the a group specificity: equal concentrations of purified 22 nm HBsAg particles of adw, ayw or adr subtype inhibited the CHId-anti-idiotype reaction to the same degree (26). Both synthetic peptides efficiently inhibited the reaction (27). The ability to react with human anti-HBs demonstrated that the epitopes present on the peptides are also expressed as immunogenic determinants on naturally produced HBV particles. The role of conformation was emphasized, in that reduction of the disulfide bond and alkylation of free thiol groups resulted in a linear peptide which no longer blocked the CHId-anti-idiotype reaction (27). Similarly, reduced and alkylated preparations of non-denatured P25 + GP30 micelles also failed to inhibit this reaction (27).

In summary the following observations were made:

1) Inoculation of SP1 induced anti-HBs antibodies in experimental animals.

2) SP1 coupled to a protein carrier (tetanus toxoid) induced in mice anti-HBs titers comparable to those elicited by intact 22 nm HBsAg particles.

3) SP1 was found to be much more antigenic than linear HBsAg synthetic peptides; this is due to the secondary structure conferred on the a epitope by the disulfide bond.

4) A conformational a and a sequential y epitope were detected on SP1.

5) The disulfide bond was essential for the _a_ activity.

6) The _a_ epitope present on SP1 appeared to be a predominant epitope recognized by humans in response to a natural HBV infection.

Thus, preparation of defined synthetic peptides may represent a feasible advance over the current HBV vaccine. The advantages would be no dependency on human HBV carriers for source of immunogenic material and no requirement for safety-testing in chimpanzees. In addition, there would be a true reproducibility of composition of a vaccine that could be produced in large supplies at a much lower cost than the present product.

REFERENCES

1. Dreesman, G. R., Hollinger, F. B., McCombs, R. M., and Melnick, J. L. (1972). Infec. Immun. 5:213.
2. Cabral, G. A., Marciano-Cabral, F., Funk, G. A., Sanchez, Y., Hollinger, F. B., Melnick, J. L., and Dreesman, G. R. (1978). J. Gen. Virol. 38:339.
3. Sanchez, Y., Ionescu-Matiu, I., Hollinger, F. B., Melnick, J. L., and Dreesman, G. R. (1980). J. Gen. Virol. 48:273.
4. Dreesman, G. R., Hollinger, F. B., Sanchez, Y., Oefinger, P., and Melnick, J. L. (1981). Infec. Immun. 32:62.
5. Dreesman, G. R., Hollinger, F. B., McCombs, R. M., and Melnick, J. L. (1973). J. Gen. Virol. 19:129.
6. Tiollais, P., Charnay, P., and Vyas, G. N. (1981). Science 213:406.
7. Sanchez, Y., Ionescu-Matiu, I., Melnick, J. L., and Dreesman, G. R. (1983). J. Med. Virol. 11:115.
8. Simons, K., Helenius, A., Leonard, K., Sarvas, M., and Gething, M. J. (1978). Proc. Natl. Acad. Sci. USA 75:5306.
9. Zuckerman, A. J., Howard, C. R., and Skelly, J. (1981). _in_ Hepatitis B Vaccine, P. Maupas and P. Guesry, eds., Elsevier/North-Holland Biomedical Press, Amsterdam, p. 251.
10. Ionescu-Matiu, I., Sanchez, Y., Fields, H. A., and Dreesman, G. R. (1983). J. Virol. Methods 6:41.
11. Charnay, P., Pourcel, C., Louise, A., Fritsch, A., and Tiollais, P. (1979). Proc. Natl. Acad. Sci. USA 76: 2222.
12. Valenzuela, P., Gray, P., Quiroga, M., Zaldivar, J., Goodman, A. M., and Rutter, W. J. (1979). Nature 280:815.
13. Pasek, M., Gotto, T., Gilbert, W., Zink, B., Schaller, H., MacKay, P., Leadbetter, G., and Murray, K. (1979). Nature 282:575.
14. Peterson, D. L. (1981). J. Biochem. 256:6975.

15. Bull, H. B. and Bresse, K. (1975). Arch. Biochem. Biophys. 161:665.
16. Chou, P. Y. and Fasman, G. D. (1978). Adv. Enzym. 47:45.
17. Lerner, R. A., Green, N., Alexander, H., Liu, F.-T., Sutcliffe, J. G., and Shinnick, T. M. (1981). Proc. Natl. Acad. Sci. USA 78:3403.
18. Bhatnagar, P. K., Papas, E., Blum, H. E., Milich, D. R., Nitecki, D., Karels, M. J., and Vyas, G. N. (1982). Proc. Natl. Acad. Sci. USA 79:4400.
19. Prince, A. M., Ikram, H., and Hopp, T. P. (1982). Proc. Natl. Acad. Sci. USA 79:579.
20. Dreesman, G. R., Sanchez, Y., Ionescu-Matiu, I., Sparrow, J. T., Six, H. R., Peterson, D. L., Hollinger, F. B., and Melnick, J. L. (1982). Nature 295:158.
21. Sanchez, Y., Ionescu-Matiu, I., Sparrow, J. T., Melnick, J. L., and Dreesman, G. R. (1982). Intervirology 18:209.
22. Kennedy, R. C., Ionescu-Matiu, I., Adler-Storthz, K., Henkel, R. D., Sanchez, Y., and Dreesman, G. R. (1983). Intervirology 19:176.
23. Ionescu-Matiu, I., Kennedy, R. C., Sparrow, J. T., Culwell, A. R., Sanchez, Y., Melnick, J. L., and Dreesman, G. R. (1983). J. Immunol. 130:1947.
24. Peterson, D. L. (1982). in Viral Hepatitis, 1981 International Symposium, W. Szmuness, H. J. Alter and J. E. Maynard, eds., Franklin Institute Press, Philadelphia, p. 707.
25. Kennedy, R. C. and Dreesman, G. R. (1983). J. Immunol. 130:385.
26. Kennedy, R. C., Sanchez, Y., Ionescu-Matiu, I., Melnick, J. L., and Dreesman, G. R. (1982). Virology 122:219.
27. Kennedy, R. C., Dreesman, G. R., Sparrow, J. T., Culwell, A. R., Sanchez, Y., Ionescu-Matiu, I., Hollinger, F. B., and Melnick, J. L. (1983). J. Virol. 46:653.

SYNTHESIS AND ASSEMBLY OF HEPATITIS B VIRUS ANTIGENS IN HETEROLOGOUS SYSTEMS

Pablo Valenzuela, Paulina Bull, Doris Coit, Brian Craine, Robert Hallewell, Ulrike Heberlein, Orgad Laub, Frank Masiarz, Angelica Medina and Steve Rosenberg

Chiron Research Laboratories
Chiron Corporation
Emeryville, California 94608

INTRODUCTION

Hepatitis B virus (HBV) is a particle of 42 nm diameter (the Dane particle) consisting of a core containing the viral genome (3200 base pairs of partially single-stranded DNA (1)) bound to the core protein and its own DNA polymerase (2,3); the virus core is surrounded by a phospholipid-containing envelope carrying the major surface antigenic determinants (HBsAg). These reside in a single protein, which occurs in both a glycosylated and non-glycosylated form of 27,000 and 23,000 daltons respectively (4,5). Infection with hepatitis B virus leads not only to the production of Dane particles but also to a dramatic overproduction of 22 nm large particles and filaments (the HBsAg particle) that contain the elements of the surface envelope. These HBsAg particles are about 1000-fold more immunogenic than the unassembled HBsAg protein (6). Core particles produced by removal of the surface coat protein contain a major polypeptide (the core antigen, HBcAg) of MW 19,000. This protein is very basic and interacts with the viral DNA in the center of the viral particle.

HBV has a very narrow host range (humans and chimpanzees) and cannot be propagated in tissue culture. Therefore, investigation of the structure and mechanism of infection of HBV has been severely restricted. The advent of molecular cloning has allowed the complete virus genome to be cloned and sequenced, and the coding regions for the different antigens have been identified (7,8,9,10), raising the possibility of expressing HBV genes in alternative host systems. This report describes our recent work on the expression and assembly of HBcAg in E. coli and of HBsAg in yeast and mammalian cells.

SYNTHESIS AND ASSEMBLY OF HEPATITIS B CORE ANTIGEN IN E. COLI

Experiments on the expression of HBcAg in heterologous systems are the first step in studies relating the structure of this protein with its function in the viral particle. In addition, HBcAg synthesized from cloned HBV DNA fragments could replace liver- or serum-isolated particles in serological assays for HBV antigens and antibodies. Several systems have been used previously for the production of hepatitis B antigens in E. coli (11,12,13). We describe here the construction of recombinant plasmids which direct the synthesis of high levels of HBcAg. The antigen found in the bacterial extracts is assembled into particles similar to the viral cores seen in infected human livers. In these plasmids, as shown in Figure 1, the hepatitis B core antigen coding region has been fused to the beginning of the β-galactosidase gene under the control of a hybrid E. coli trp promoter and E. coli lac operator. A fragment of 1005 bp containing the region coding for the HBcAg was isolated by gel electrophoresis after digesting the plasmid pHBV3200 (7) with Hha1. This fragment was then treated with Bal-31 exonuclease, ligated to Pst1 oligonucleotide linkers, and ligated to plot4 which had been previously digested with pst1. The resulting

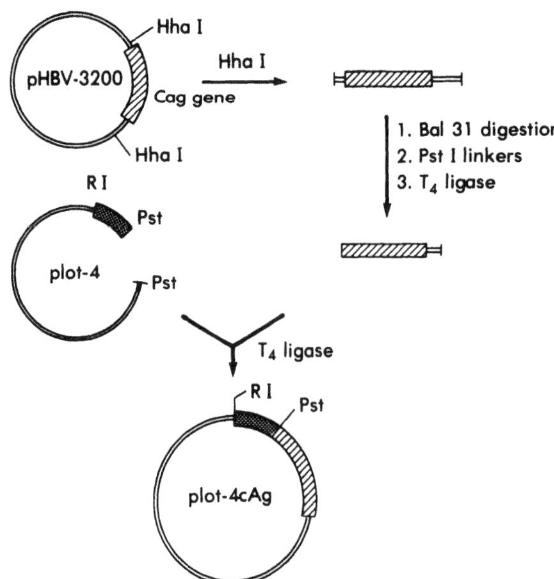

Figure 1. Construction of plasmids for the expression of HBcAg in E. coli. The details are described in the text.

collection of plasmids (plot4cAg) contains the region coding for HBcAg fused to the first seven amino acids of the lacZ gene, half of them in the proper orientation and one third of those in the proper reading frame. Colonies obtained after transformation of E. coli with plot4cAg were analyzed for the expression of HBcAg by a solid phase radioimmunoassay carried out on bacterial colonies. Positive colonies were grown in liquid cultures, and extracts were analyzed for the amounts of HBcAg by a competitive radioimmunoassay using the Abbott HBcAg antibody detection kit. Extracts from the most active E. coli transformants competed 50% of the core antigen binding after a 1:1000 dilution and could be easily detected after a 1:10,000 dilution. Extracts from plasmid-free E. coli cells did not have RIA activity.

Preliminary gel filtration experiments indicated that the HBcAg synthesized in E. coli has a molecular weight higher than that of the expected monomer protein. Centrifugation to equilibrium in cesium chloride density gradients indicated that the HBcAg synthesized in E. coli bands at approximately 1.35 g/ml, which suggests that they contain nucleic acid. The nature of the material was analyzed further by electron microscopy after staining the peak fractions of the density gradient with 2% uranyl acetate. The results, which are shown in Figure 2, indicate that these fractions contain small round particles of about 25 to 27 nm of diameter. A similar result has recently been reported by Cohen and Richmond (14). Recent experiments (data not shown) indicate that these particles are formed by the assembly of core protein with plasmid DNA (A. Medina, P. Valenzuela, unpublished results). Preliminary experiments (results not shown) indicate that the HBcAg synthesized in E. coli is as specific and sensitive as the liver-derived antigen in serological assays for hepatitis B.

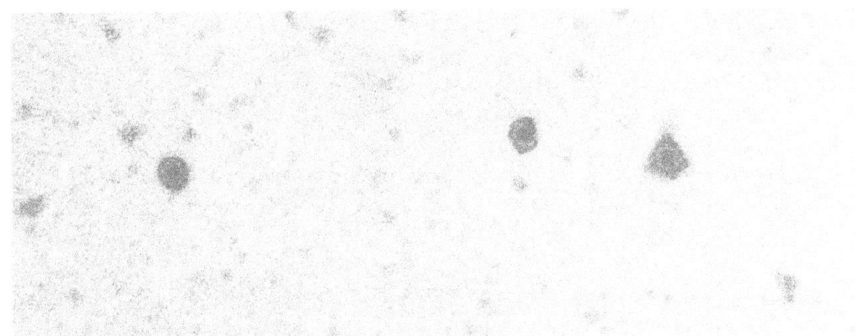

Figure 2. Electron micrographs of HBcAg particles produced by E. coli. Particles were purified by CsCl density centrifugation and visualized after negative staining with 2% uranyl acetate.

SYNTHESIS AND ASSEMBLY OF HEPATITIS B SURFACE ANTIGEN IN YEAST

Yeast is an attractive alternative host system. This organism has complex membrane systems and the ability to glycosylate and secrete proteins by processes resembling those observed in higher cells (15). In addition, yeast has also been developed as a system which allows the introduction, maintenance and expression of foreign genes (16,17).

For the efficient expression of the HBsAg in yeast, we have employed the yeast glyceraldehyde 3P-dehydrogenase (GAPDH) promoter instead of the yeast alcohol dehydrogenase 1 promoter used previously (18). GAPDH is produced at high levels in yeast under certain nutritional circumstances (growth on glucose, low oxygen), and there is evidence which suggests that this gene is regulated at the level of transcription (19).

The isolation of the GAPDH gene used in these studies has been reported elsewhere (20). In short, we have isolated the GAPDH gene as a 3200 bp BamH1 fragment cloned in pBR322 (pGAP-2). This fragment contains 1500 bp of coding region, 1200 bp of 5' flanking region and 500 bp of 3' flanking sequences. The construction of the yeast expression plasmid pHBS-70 is shown in Figure 3. DNA from pGAP-2 was digested with Xbal, partially digested with Bal-31 exonuclease and finally digested with Ncol. The resulting linear molecules were ligated to the HBsAg-coding sequence, which was isolated from pHBS-5-3, a plasmid containing the HBsAg gene cloned as a EcoR1-Hind111 fragment in pBR322. This fragment contains an Ncol site on bp preceding the ATG which encodes the N-terminal methionine of mature HBsAg. After ligation at the Ncol ends, the other ends of the molecules were filled in with deoxynucleotides and DNA polymerase, ligated and used to transform _E. coli_. Plasmid pGAP-HBS was obtained which contains a fusion containing the GAPDH promoter region, the HBsAg gene and part of GAPDH coding and termination region in the proper orientation. This "cassette" was excised by digesting pGAP-HBS DNA with BamH1 and ligating it to BamH1 digested pCH-1. This plasmid vector contains yeast 2 micron sequences including a replication origin the yeast leu_2 gene for selection in yeast cells and part of pBR322 including the replication origin and the ampicillin resistance gene for selection in _E. coli_. The resulting plasmid pHBS-70 was obtained.

Yeast cells (strain 2150-2-3) were transformed with pHBS-70 and cell extracts from mid-log phase were assayed for HBsAg by the AUSRIA II radioimmunoassay. A substantial level of HBsAg was detected in cells transformed with pHBS-70, and no antigen was observed in cells transformed with the parent plasmid pCh-1. The amount of protein made per 200 ml of yeast culture is about 50 to 100 ug.

The predominant form of HBsAg produced in human cells is the so-called 22 nm particle. Its biophysical properties are well documented (21,22), and its immunological potency exceeds that of the pure protein (6). To examine the form in which surface antigen is present in yeast, we subjected extracts to equilibrium sedimentation through a discontinuous CsCl gradient. A control tube containing HBsAg from the human hepatoma cell line PLC/PRF/5 (23) was treated identically to provide a buoyant density marker. HBsAg synthesized by yeast was found to band at the same density as that from the PLC/PRF/5 cell, that is, 1.2 g/ml. Peak fractions of the CsCl gradient were analyzed further by velocity sedimentation in sucrose gradients. Again the peak of HBsAg produced by yeast coincided exactly with HBsAg from human cells which sediments at approximately 55S (22).

From the sedimentation data, it is apparent that the HBsAg synthesized in yeast under the control of the yeast GAPDH promoter is in the form of particles or aggregates. The nature of

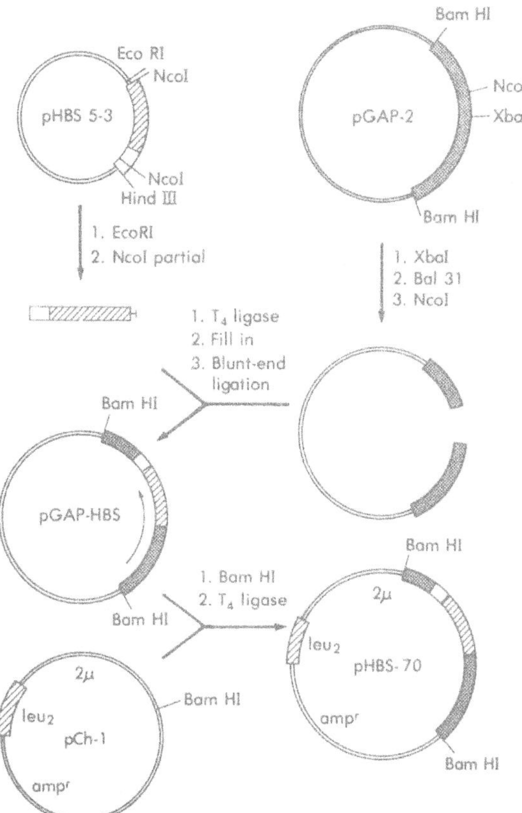

Figure 3. Construction of plasmids for the expression of HBsAg in yeast using the yeast glyceraldehyde 3-P dehydrogenase promoter. The details are described in the text.

these particles was further characterized by electron microscopy. HBsAg synthesized in yeast was purified by affinity chromatography and negatively stained with uranyl acetate. When examined under the electron microscope, the yeast HBsAg preparation was found to contain particles (Figure 4) with the shape of short filaments 20 nm wide and 50 nm long. Preliminary tests have shown that these particles, as those obtained previously with the ADH1 promoter (18), are antigenic, inducing antibodies in animal species (G. Buynak and W. McAleer, personal communication).

The chemical composition of HBsAg molecules synthesized in yeast was determined by immunoprecipitation of ^{35}S-proteins labeled in vivo in a yeast strain transformed by the GAPDH-HBsAg recombinant plasmid. SDS gel electrophoresis of the anti-HBsAg immunoreactive material revealed a single band of apparent MW 25,000 corresponding in size to the unglycosylated HBsAg component of PLC/PRF/5 cells (results not shown). No higher molecular weight bands could be detected, indicating that most of the molecules were not glycosylated in these conditions.

In conclusion, expression of hepatitis B surface antigen coding sequences in yeast leads to the production of particles

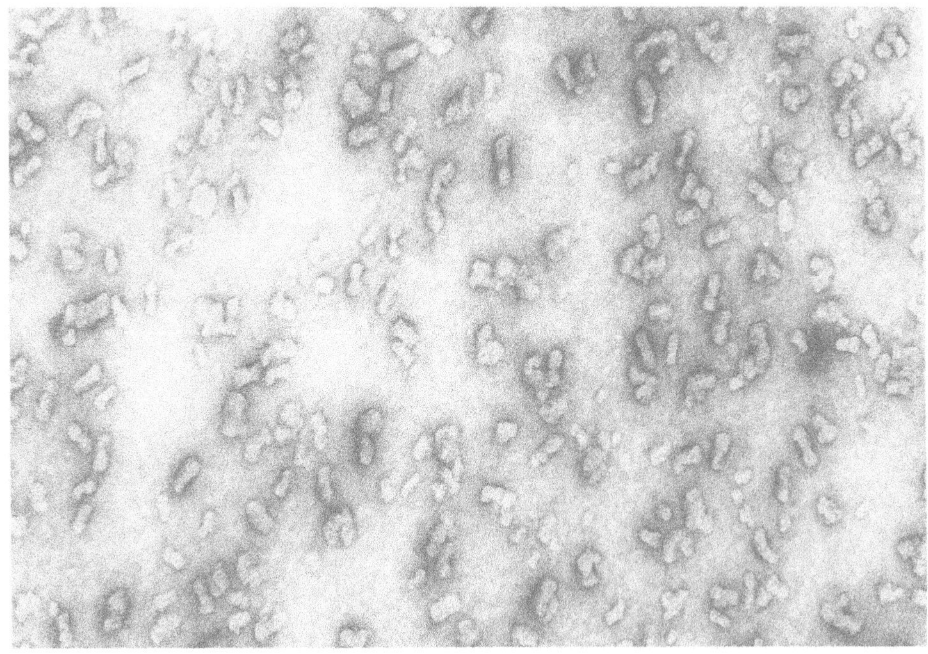

Figure 4. Electron micrographs of HBsAg particles produced by yeast. Particles were purified by immunoaffinity chromatography and visualized after negative staining with 2% uranyl acetate.

which react with anti-HBsAg antibodies. These particles are similar to those made by human carrier patients or by a hepatoma cell line; they have identical sedimentation rates and buoyant density, suggesting a similar size and ratio of protein to lipid composition for yeast and for human particles. However, unlike the HBsAg from human cells, the yeast particle does not contain significant quantities of a higher molecular weight glycoprotein, indicating that glycosylation is required neither for the formation of the particulate structures nor for immunogenicity.

The construction we have used in the HBsAg-expressing plasmid eliminates the region of the gene encoding the leader peptide of putative pre-HBsAg. Therefore, formation of particles in yeast implies that the polypeptide sequence of the surface antigen molecule contains the requisite instructions for assembly of apparently normal particles within the cytoplasmid milieu of the yeast cell. The putative amino terminal leader region and other possible precursors are not required for the formation of this structure. Also, no other hepatitis gene products or liver-specific cell products are necessary to form the particulate structure. Moreover, it seems likely that such particles can be formed in other heterologous cell systems, including bacteria. It remains to be determined where within the yeast cell the 22 nm particle is assembled and what biochemical processes accompany it. An immediate advantage of the expression of HBsAg in yeast is that the physiological and genetic requirements for particle assembly may be systematically explored, making use of yeast mutants conditionally defective for the secretory pathway and for other aspects of cellular metabolism.

The similarity in structure of the yeast particle to bona fide 22 nm particles and the high immunogenicity in animals emphasizes the possible value of the HBsAg particle as a vaccine. The yeast HBsAg particles can in principle be produced in large quantities for this purpose. Further, the complete absence of 42 nm Dane particles, HBV DNA and/or human proteins in such preparations eliminates the possibility of secondary infections or autoimmunity problems.

SYNTHESIS AND ASSEMBLY OF HEPATITIS B SURFACE ANTIGEN IN MAMMALIAN CELLS

For our studies on the synthesis and assembly of the HBsAg in mammalian cells, we have constructed expression vectors derived from SV40. These recombinants retain the SV40 late genes, the early promoter and polyadenylation site, but heterologous sequences are substituted for the coding region of T antigen. The latter function is constitutively produced by the host SV40 transformed monkey kidney cells (COS) cells (24) and, therefore, efficient replication and subsequent packaging of these molecules occurs in this system. All early SV40 splice junctions have been deleted from the SV40 vector, thus aberrant splicing caused by these sequences is eliminated. Further, the relatively strong SV40 early promoters and the high number of gene copies per cell achieved in this system result in the formation of high cellular levels of recombinant RNA.

The construction of the SV40-HBsAg recombinant used in this study is summarized in Figure 5. The construction is based on the fact that the early genes of the SV40 DNA are flanked by sites for the restriction endonucleases Hind111 and Bc11. The Hind111 site is located at nucleotide number 5171 on the SV40 genome and is 8 nucleotides 5' to the initiation codon of the large T antigen gene. The Bc11 site is located at nucleotide number 2770 which is 77 nucleotides 5' to the termination codon of the large T antigen gene. The SV40 genome cloned in the BamH1 site of pBR322 (pSV40) was amplified in E. coli strain GM48. pSV40 DNA was linearized by partial digestion with Hind111 followed by S1 treatment to produce blunt ends. The linear pSV40 DNA was digested with Bc11, and the

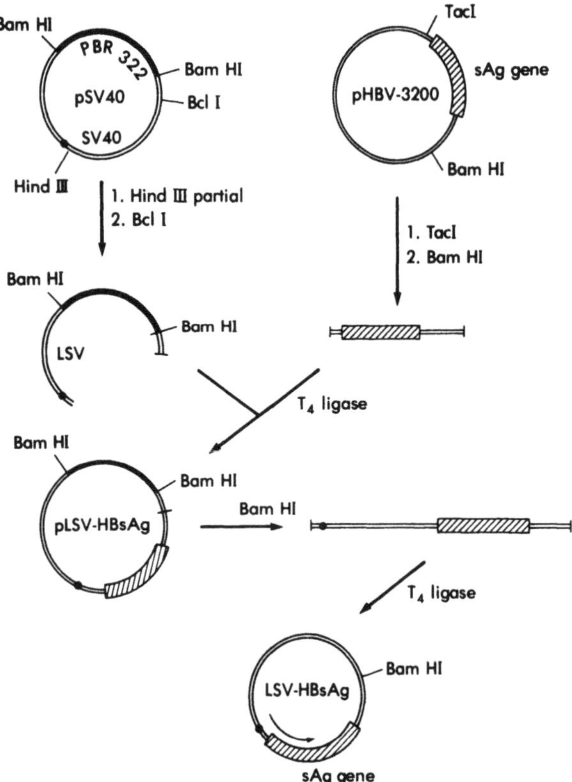

Figure 5. Construction of SV40-derived vector for the expression of HBsAg in mammalian cells. The details are described in the text.

7.2 kb fragment (pLSV) was purified by preparative agarose gel electrophoresis. The coding region for HBsAg was purified as a Tac1-BamH1 fragment (nucleotides 130-1403 on the cloned HBV genome (10)). This fragment was inserted in the pLSV vector, and ampicillin resistant colonies were screened by hybridization to ^{32}P-labeled SV40 and HBV probes. The resulting plasmids contain an SV40 origin of replication, a functional set of SV40 late genes and HBsAg cloned in the correct orientation with respect to the SV40 early promoter and the early genes polyadenylation site. The BamH1 fragment containing the LSV-HBV recombinants was isolated, self-ligated to form circular DNA and transfected into COS cells that produce T antigen and are hence permissive for SV40 early replacement recombinants (24). Viruses containing the LSV-HBV DNA were efficiently propagated in this host, and high titer stocks (10^8 pfu/ml) were obtained following a second round of viral production. Restriction analysis of the purified viral DNA showed that no rearrangements had occurred during replication.

COS cells infected with the LSV-HBsAg recombinant secrete up to 800 ng of HBsAg per 10^6 cells, an equivalent of over 10^7 molecules per cell. The purified HBsAg particles are stable and have been stored for several months without any loss in immunoreactivity. In addition, large quantities (5-10 ug/ml) of HBsAg accumulate if the viral infection is allowed to progress for 10 days.

The HBsAg molecules produced by the SV40-COS cell system were characterized by immunoprecipitation of ^{35}S-labeled proteins. SDS gel electrophoresis of the anti-HBsAg immunoreactive material disclosed three specific bands (results not shown). The two predominant peptides, molecular weight 23,000 and 27,000, correspond to the nonglycosylated and glycosylated forms of HBsAg, respectively. The third band, about 46,000 in molecular weight, may be a HBsAg dimer (results not shown).

The nature of the secreted HBsAg particles was further characterized by electron microscopy. Affinity purified HBsAg coded by LSV-HBsAg was absorbed onto carbon film grids and negatively stained with uranyl acetate. When examined under the electron microscope, the HBsAg preparation revealed a homogeneous population of spherical 22 nm particles (Figure 6) which are similar to the particles found in the sera of infected individuals.

Further, electron microscopic analysis showed that the secreted HBsAg particles coded by the LSV-HBsAg recombinant are similar in structure to the 22 nm particles detected in the sera of infected individuals.

The efficient production of HBsAg by the LSV-HBsAg recombinant indicates that the putative presurface signal peptide

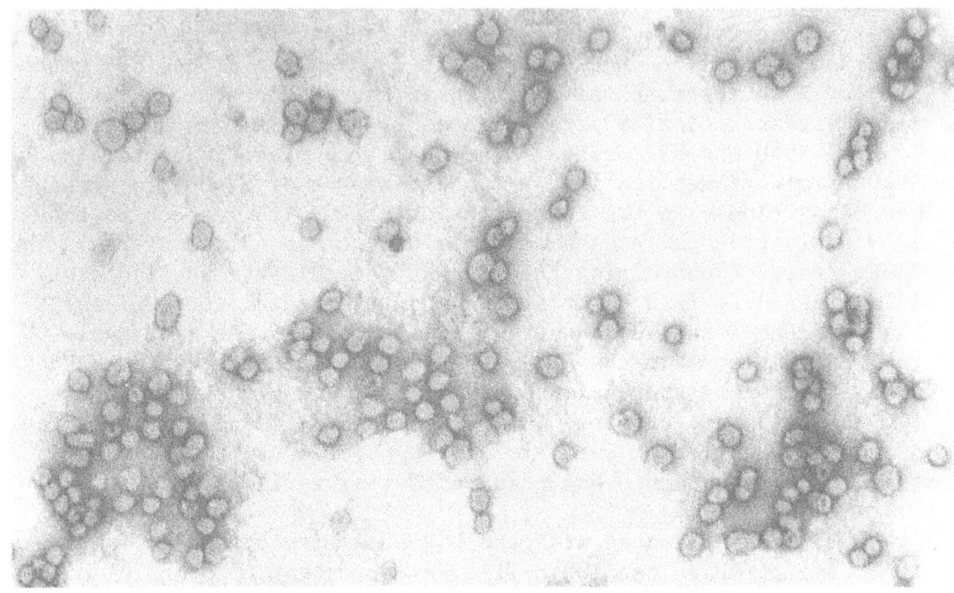

Figure 6. Electron micrographs of HBsAg particles produced by transfected COS cells in culture. Particles were purified by immunoaffinity chromatography and visualized after negative staining with 2% uranyl acetate.

preceding the coding region of mature HBsAg is not required for the assembly, secretion and stability of the HBsAg particles.

The amount of HBsAg secreted by COS cells infected with the LSV-HBsAg recombinant is substantial, about 10^7 HBsAg molecules per infected cell during the 24-hour period prior to the initiation of cell lysis. This is about 10 fold more than that produced by the human PLC/PRF/5 cell line or by several HBV/SV40 late replacement recombinants studied elsewhere (24,25).

SUMMARY

Molecular cloning and nucleotide sequence of the hepatitis B DNA has revealed the organization of the viral genome and the amino acid sequence of the two structural proteins of the virus, the core antigen and the surface antigen. Using this information, it has been possible to design and construct recombinant DNA molecules directing the expression of these antigens in alternative hosts.

E. coli cells transformed with recombinant plasmids containing the core antigen coding region properly fused to a bacterial promoter system synthesize and assemble core antigen spherical particles of approximately 27 nm in diameter.

The surface antigen coding region has been inserted in front of yeast promoters in plasmids capable of autonomous replication. Yeast cells transformed with these molecules synthesize and assemble surface antigen particles of size and structure similar to the particles produced in liver during viral infection. Recombinant SV40 DNA molecules in which the surface antigen coding region has been inserted under the control of the early SV40 promoter have been constructed. Monkey kidney cells transfected with these molecules synthesize, assemble and secrete glycosylated 22 nm surface antigen particles.

These studies demonstrate the potential use of genetically engineered cells in the synthesis of viral antigens for diagnostic and immunization purposes.

ACKNOWLEDGMENTS

We thank Graeme Bell for his helpful comments on the manuscript and friends and scientists at Chiron Corporation for comments and support.

REFERENCES

1. Summers, J., O'Connell, A., and Millman, I. (1975). Proc. Natl. Acad. Sci. USA 72:4597.
2. Robinson, W. S. and Greenman, R. C. (1974). J. Virol. 13: 1231.
3. Hruska, J. F. and Robinson, W. S. (1977). J. Med. Virol. 1: 119.
4. Shih, J. W. K. and Gerin, J. L. (1977). J. Virol. 21:1219.
5. Peterson, D. L. (1981). J. Biol. Chem. 256:6975.
6. Cabral, G. A., Marciano-Cabral, F., Funk, G. A., Sanchez, Y., Hollinger, F. B., Melnick, J. L., and Dreesman, G. R. (1978). J. Gen. Virol. 38:339.
7. Valenzuela, P., Gray, P., Quiroga, M., Zaldivar, J., Goodman, H. M., and Rutter, W. J. (1979). Nature 280:815.
8. Pasek, M., Goto, T., Gilbert, W., Zink, B., Schaller, H., Mackay, P., Leadbetter, G., and Murray, K. (1979). Nature 282:575.
9. Galibert, F., Mandart, E., Fitoussi, F., Tiollais, P., and Charnay, P. (1979). Nature 281:646.

10. Valenzuela, P., Quiroga, M., Zaldivar, J., Gray, P., and Rutter, W. J. (1980). in Animal Virus Genetics, B. N. Fields, R. Jaemisch and C. F. Fox, eds., Academic Press, New York, pp. 55-77.
11. Burrell, C. J., Mackay, P., Greenway, P. J., Hofschneider, P. H., and Murray, K. (1979). Nature 279:43.
12. Edman, J. C., Hallewell, R. A., Valenzuela, P., Goodman, H. M., and Rutter, W. J. (1981). Nature 291:503.
13. Stahl, S., Mackay, P., Magazin, M., Bruce, S. A., and Murray, K. (1982). Proc. Natl. Acad. Sci. USA 79: 1606.
14. Cohen, B. and Richmond, J. E. (1982). Nature 296:677.
15. Novick, P., Field, C., and Scheckman, R. (1980). Cell 21: 205.
16. Beggs, G. D. (1978). Nature 275:104.
17. Petes, P. D. (1980). Ann. Rev. Biochem. 49:845.
18. Valenzuela, P., Medina, A., Rutter, W. J., Ammerer, G., and Hall, B. D. (1982). Nature 298:347.
19. Maitra, P. K. and Lobo, Z. (1971). J. Biol. Chem. 246:489.
20. Urdea, M. S., Merryweather, J. P., Mullenbach, G. T., Coit, D., Heberlein, U., Valenzuela, P., and Barr, P. J. (Submitted for publication.)
21. Gerin, J. L., Purcell, R. H., Hoggan, M. D., Holland, P. V., and Chanock, R. M. (1969). J. Virol. 4:763.
22. Gerin, J. L., Holland, P. V., and Purcell, R. H. (1971). Virology 7:569.
23. Macnab, G. M., Alexander, J. J., Letcasas, G., Bey, E. M., and Urbanowicz, J. M. (1976). Br. J. Cancer 34: 509.
24. Gluzman, Y. (1981). Cell 23:175.
25. Moriarty, A. M., Hoyer, B. H., Shih, J. W., Gerin, J. L., and Hammer, D. H. (1981). Proc. Natl. Acad. Sci. USA 78:2606.
26. Liu, C. C., Yausura, D., and Levinson, A. D. (1982). DNA 1:213.

CLOSING REMARKS

Irving Millman

Institute for Cancer Research
Fox Chase Cancer Center
Philadelphia, Pennsylvania 19111

We have now come to the conclusion of this meeting on the subject of hepatitis B; the virus, the disease and the vaccine. I will try to sum up briefly what I believe has been the substance of this meeting.

Dr. Blumberg has given us a clear picture of events from as early as 1963 which led to the discovery of the hepatitis B virus, the present vaccine and the Nobel Prize. His current activities are dealing with the mechanism by which the hepatitis virus is involved in the pathogenesis of primary hepatic carcinoma.

Dr. Summers has revealed the life cycle of the hepatitis B virus. The DNA genome is composed of a complete minus strand with a protein covalently linked to its 5' end, duplexed with an incomplete or nascent plus strand. These DNA genomes have been found to be synthesized by reverse-transcription of a plus strand RNA pre-genome, a mechanism which is strikingly similar to the retroviruses.

Dr. London reported that in Asia and Africa 40-90% of the patients who have chronic active hepatitis, cirrhosis or PHC are chronic carriers compared with 5-15% of controls. Forty to seventy percent of mothers of patients with PHC are chronic carriers. Liver tissues from chronic hepatitis patients frequently contain HBsAg and HBV DNA. Prospective studies in Japan and Taiwan indicate that almost all cases of PHC occur in chronic carriers and the relative risk for carriers to develop PHC is 350 times that of non-carriers.

Dr. Dienstag reported that there are 200 million HBV carriers which means that 5% of the earth's population is infected. He discussed the geographic distribution and usual routes of transmission as well as population groups with enhanced risks of exposure, and the susceptibility of contacts. In nonendemic countries, transmission is via percutaneous inoculation and intimate contact, especially sexual. In developing countries, perinatal transmission is the most common mode.

Dr. Rubin described the different types of hepatitis from A to non-A, non-B (NANB), the propensity of NANB for liver disease, and complications of hepatitis B and treatments.

Dr. Ling described the variety of diagnostic tests available and the interpretation of results.

Dr. Werner reported on the efficacy and immunogenicity of the vaccine in health care personnel in six Boston hospitals where 2117 sera were screened to determine eligibility for participation in vaccine trials. As a result, 1330 received either vaccine or placebo. Screening and follow-up revealed different immunological responses including anamnestic and borderline levels of serological markers which were difficult to evaluate. She supports the adoption of an anti-HBs assay as a screening test prior to vaccination.

Dr. Schneider ended the Thursday session with remarks concerning the hazards of hepatitis B infections in dialysis and hospital ward personnel and dentists.

Dr. Millman began the Friday session by describing the early events which led to the concept and development of the vaccine.

Dr. Stevens described the two clinical trials, one involving homosexual men and the other dialysis staff. Both populations showed a similar frequency of side effects, similar antibody responses, and virtually identical efficacy. The protective efficacy of the vaccine was 92% in both trials. The vaccine, composed only of "ad" particles, protected against infection with "ay" virus. Protection by vaccination persisted for up to 3-3½ years.

Dr. Weibel confirmed the trial data of Stevens and indicated that newborn infants and children to 10 years of age showed vigorous antibody responses with three 10 ug doses of vaccine at 0, 1 and 6 months.

Dr. Bernier discussed the vaccine's high cost, limited availability, and recommendations for its use in the USA. He stated that a broad vaccination policy is not currently cost effective at this time.

Next a panel discussion followed on the topic of "Vaccine Economics: Issues and Answers." Dr. McLean from Merck discussed the production procedure for the current vaccine, Drs. Brightman and Weibel the problems encountered among professional personnel such as dentists where dental patients may be at risk from HBsAg carrier dental personnel. Dr. Gerety discussed some of the FDA requirements for manufacture of the vaccine. Less than one acceptable vaccine plasma donor can be found among 10,000 normal donors; as many as 80 can be found among members of certain groups at high risk for hepatitis B. In addition, he related the safety and efficacy requirements for each lot of vaccine produced, such as sterility testing, tissue culture inoculation to rule out adventitious viruses, and immunogenicity tests in mice to compare new lots with a standard vaccine. A total of 22 doses of any lot are injected i.v. into susceptible chimpanzees. To date more than 15 lots have been tested with no evidence of infectious virus.

The final group of sessions involved second and third generation vaccines from a molecular biology point of view. Since various studies suggested that the nature of HBV transcription depended on the integration state of viral DNA, Dr. Chakraborty found it relevant to examine RNA expression of the HBV genome in chimpanzee liver compared with the PLC/PRF/5 cell line because specific HBV transcripts might represent unique RNA products involved in various aspects of viral infection and/or tumor biogenesis. The chimpanzee liver contained two poly A RNA transcripts, a 19S and a 24S. In the liver cell line he detected a 19S transcript which was different from the 19S of the chimpanzee liver, since the former contained coding sequences for both surface and core antigen whereas the latter contained coding sequences for only surface antigen.

Dr. Ionescu-Matiu discussed the immunogenic activities of two alternative HBV vaccine preparations, a P25-GP30 polypeptide and glycopeptide derived from purified 22 nm particles and a small synthetic HBsAg peptide all tested in BALB/c mice. The first polypeptide, made by two different procedures, was compared with intact HBsAg. One of the preparations, called nondenatured micelles, reportedly had better immunoreactivity than intact HBsAg. The synthetic preparation, SP1, was a peptide containing amino acids 122-137 from P25, and was said to produce anti-HBs in 80-100% of the inoculated mice after two injections of alum precipitated material either in micelle form or covalently coupled to tetanous toxoid. SP1 contains an "a" conformation-dependent epitope and a sequential y epitope.

Molecular cloning and nucleotide sequence of the hepatitis B DNA has revealed the organization of the viral genome and the amino acid sequence of the two structural proteins of the virus, the core and surface antigen. Dr. Valenzuela used this information to design and construct recombinant DNA molecules directing the expression of these antigens in alternative hosts. The surface antigen coding region was inserted in front of a variety of yeast promoters in plasmids capable of autonomous replication in yeast. These yeast cells synthesize and assemble surface antigen particles of size and structure similar to the particles produced in liver during viral infection. Recombinant SV 40 DNA molecules in which the surface antigen coding region had been inserted under the control of the early SV 40 promoter were also constructed. Monkey kidney cells transfected with these molecules synthesize, assemble and secrete glycosylated 22 nm surface antigen particles.

I wish to thank all the participants for an excellent symposium which was highly informative. I wish to thank everyone on the Committee for a job well done. I especially wish to thank Merck & Company for their generous contribution without which this symposium would not have been possible. If there is no objection, I'd like to call this meeting to an end.

ABBREVIATIONS

AFP: Alpha fetoprotein. A protein which is normally present in fetal blood and disappears at birth, but frequently reappears in the serum of patients with primary hepatocellular carcinoma.

AIDS: Acquired immune deficiency syndrome.

ALT: L-alanine:2-oxoglutarate aminotransferase, called alanyl or alanine aminotransferase in the application and frequently abbreviated SGPT. An enzyme normally found in the cytoplasm of hepatocytes. Increased concentrations in serum indicate liver cell injury.

Anti-HBc: Antibody to hepatitis B core antigen.

Anti-HBe: Antibody to the e antigen associated with hepatitis B virus.

Anti-HBs: Antibody to hepatitis B surface antigen.

Au: Australia antigen (see HBsAg).

Ausab: A radioimmunoassay for anti-HBs (Abbott Labs).

Auszyme: An enzyme immunoassay (Abbott Labs) for detection of HBsAg and, by cross-reactivity, WHsAg.

CAH: Chronic active hepatitis.

CEP: Counterelectrophoresis. A quick, but relatively non-specific and less sensitive, method used for detection of several proteins in serum (e.g., HBsAg, HBcAg, AFP).

CLD: Chronic liver disease.

Con A: The lectin concanavalin A obtained from jackbeans (*Canavalis ensiformis*).

Corab: A radioimmunoassay for anti-HBc (Abbott Labs).

DHBV: Duck hepatitis B virus.

DNA: DNA polymerase. Used to refer to the endogenous enzyme in HBV and related viruses.

EIA: Enzyme linked immunoassay (often called ELISA). A sensitive method to assay proteins quantitatively and qualitatively.

GSHV: Ground squirrel hepatitis virus.

HBcAg: Hepatitis B core antigen. An antigen on the core of the Dane particle.

HBeAg: Originally called "e" antigen, a soluble protein ($\sim$ 19,000 daltons) present in the blood of some individuals with hepatitis B surface antigen. It is derived from the core of the hepatitis B virion (Dane particle) and is an excellent marker of infectivity.

HBsAg: Hepatitis B surface antigen, originally called Australia antigen. It is the surface coat of the Dane particle and is also present in serum in smaller particulate forms.

HBV: Hepatitis B virus.

Heptavax-B: Hepatitis B vaccine (Merck).

HLA: The major histocompatibility complex in man. Often refers to antigens on the cell surface encoded by this locus.

Icron: Name applied to viruses belonging to the same class as hepatitis B virus. The acronym derived from the Institute for Cancer Research (ICR) and virion.

ID: Immunodiffusion. A highly specific but relatively insensitive serologic method for detecting antigens and antibodies.

Lectin: A protein or glycoprotein which selectively binds to specific saccharides or polysaccharides.

MLE: Maximum likelihood estimate.

PAGE: Polyacrylamide gel electrophoresis.

PHC: Primary hepatocellular carcinoma.

RIA: Radioimmunoassay.

RPHA: Reverse passive hemagglutination.

SDS: The anionic detergent sodium dodecyl sulfate.

SDS-PAGE: Polyacrylamide gel electrophoresis performed with sodium dodecyl sulfate in the gel.

SGPT: Originally called serum glutamic pyruvic transaminase; now designated serum alanine aminotransferase (see ALT).

SGOT: Serum glutamic oxaloacetic transaminase.

SSc: A buffer made up of sodium chloride and sodium citrate for DNA.

WHcAg: Woodchuck hepatitis core antigen.

WHeAg: Woodchuck hepatitis e antigen.

WHsAg: Woodchuck hepatitis surface antigen.

WHV: Woodchuck hepatitis virus.

CONTRIBUTORS

Bernier, Roger H., Epidemiologic Research Section, Surveillance, Investigations and Research Branch, Division of Immunization, CDC, Atlanta, Georgia 30333, USA.

Blumberg, Baruch S., Institute for Cancer Research, Fox Chase Cancer Center, Philadelphia, Pennsylvania 19111, USA.

Brightman, Vernon J. University of Pennsylvania School of Dental Medicine, Philadelphia, Pennsylvania 19104, USA.

Bull, Paulina, Chiron Research Laboratories, Chiron Corporation, Emeryville, California 94608, USA.

Buynak, Eugene B., Virus and Cell Biology Research Department, Merck Sharp and Dohme Research Laboratories, West Point, Pennsylvania 19486, USA.

Chakraborty, Prasanta R., Department of Microbiology and Immunology, The Medical College of Pennsylvania, Philadelphia, Pennsylvania 19129, USA.

Coit, Doris, Chiron Research Laboratories, Chiron Corporation, Emeryville, California 94608, USA.

Craine, Brian, Chiron Research Laboratories, Chiron Corporation, Emeryville, California 94608, USA.

Craven, Donald E., Boston University Hospital, Boston City Hospital and Boston University School of Medicine, Boston, Massachusetts 02118, USA.

Crumpacker, Clyde S., Beth Israel Hospital and Harvard Medical School, Boston, Massachusetts 02115, USA.

Dreesman, G. R., Department of Virology and Epidemiology, Baylor College of Medicine, Houston, Texas 77030, USA.

Dienstag, Jules L., Gastrointestinal Unit (Medical Services), Massachusetts General Hospital and Department of Medicine, Harvard Medical School, Boston, Massachusetts 02115, USA.

Eisenstein, Toby K., Temple University School of Medicine, Philadelphia, Pennsylvania 19140, USA.

Francis, Donald P., Hepatitis Branch, Center for Infectious Diseases, CDC, Atlanta, Georgia 30333, USA.

Gerety, Robert J., Office of Biologics, National Center for Drugs and Biologics, Food and Drug Administration, Bethesda, Maryland 20205, USA.

Grady, G. F., State Laboratory Institute, Jamaica Plain, Massachusetts, Massachusetts Department of Public Health and Tufts University School of Medicine, Boston, Massachusetts 02130, USA.

Hallewell, Robert, Chiron Research Laboratories, Chiron Corporation, Emeryville, California 94608, USA.

Heberlein, Ulrike, Chiron Research Laboratories, Chiron Corporation, Emeryville, California 94608, USA.

Hilleman, Maurice R., Virus and Cell Biology Research Department, Merck Sharp and Dohme Research Laboratories, West Point, Pennsylvania 19486, USA.

Ionescu-Matiu, I., Department of Virology and Epidemiology, Baylor College of Medicine, Houston, Texas 77030, USA.

Kane, Mark A., Hepatitis Branch, Center for Infectious Diseases, CDC, Atlanta, Georgia 30333, USA.

Kennedy, R. C., Department of Virology and Epidemiology, Baylor College of Medicine, Houston, Texas 77030, USA.

Kuter, Barbara J., Virus and Cell Biology Research Department, Merck Sharp and Dohme Research Laboratories, West Point, Pennsylvania 19486, USA.

Laub, Orgad, Chiron Research Laboratories, Chiron Corporation, Emeryville, California 94608, USA.

Ling, Chung-Mei, Abbott Laboratories, North Chicago, Illinois 60064, USA.

London, W. Thomas, Institute for Cancer Research, Fox Chase Cancer Center, Philadelphia, Pennsylvania 19111, USA.

Masiarz, Frank, Chiron Research Laboratories, Chiron Corporation, Emeryville, California 94608, USA.

McLean, Arlene A., Virus and Cell Biology Research Department, Merck Sharp and Dohme Research Laboratories, West Point, Pennsylvania 19486, USA.

Medina, Angelica, Chiron Research Laboratories, Chiron Corporation, Emeryville, California 94608, USA.

Melnick, J. L., Department of Virology and Epidemiology, Baylor College of Medicine, Houston, Texas 77030, USA.

Millman, Irving, Institute for Cancer Research, Fox Chase Cancer Center, Philadelphia, Pennsylvania 19111, USA.

Nathanson, Neal, Department of Microbiology, University of Pennsylvania School of Medicine, Philadelphia, Pennsylvania 19104, USA.

Platt, Richard, New England Deaconess Hospital and Brigham and Women's Hospital, Boston, Massachusetts 02115, USA.

Polk, B. Frank, Channing Laboratory and Harvard Medical School, Boston, Massachusetts 02115, USA.

Rosenberg, Steve, Chiron Research Laboratories, Chiron Corporation, Emeryville, California 94608, USA.

Rubin, Emanuel, Department of Pathology and Laboratory Medicine, Hahnemann University School of Medicine, Philadelphia, Pennsylvania 19102, USA.

Ruiz-Opazo, Nelson, Department of Pediatrics/Cardiology, Children's Hospital Medical Center, Harvard Medical School, Boston, Massachusetts 02115, USA.

Sanchez, Y., Department of Virology and Epidemiology, Baylor College of Medicine, Houston, Texas 77030, USA.

Schneider, William J., Albert Einstein College of Medicine and Montefiore Hospital, Bronx, New York 10461, USA.

Shafritz, David A., Departments of Medicine/Cell Biology, Albert Einstein College of Medicine, Bronx, New York 10461, USA.

Skarinsky, Evgenya, Department of Pathology and Laboratory Medicine, Hahnemann University School of Medicine, Philadelphia, Pennsylvania 19102, USA.

Snydman, David R., Tufts-New England Medical Center and Tufts University School of Medicine, Boston, Massachusetts 02130, USA.

Sparrow, J. T., Department of Virology and Epidemiology, Baylor College of Medicine, Houston, Texas 77030, USA.

Summers, Jesse, Institute for Cancer Research, Fox Chase Cancer Center, Philadelphia, Pennsylvania 19111, USA.

Valenzuela, Pablo, Chiron Research Laboratories, Chiron Corporation, Emeryville, California 94608, USA.

Weibel, Robert E., The Joseph Stokes, Jr. Research Institute, Children's Hospital of Philadelphia, Philadelphia, Pennsylvania 19104, USA.

Werner, Barbara G., State Laboratory Institute, Jamaica Plain, Massachusetts, Massachusetts Department of Public Health and Tufts University School of Medicine, Boston, Massachusetts 02130, USA.

West, David J., Virus and Cell Biology Research Department, Merck Sharp and Dohme Research Laboratories, West Point, Pennsylvania 19486, USA.

INDEX

www.ingramcontent.com/pod-product-compliance
Ingram Content Group UK Ltd.
Pitfield, Milton Keynes, MK11 3LW, UK
UKHW012204240726
13966UKWH00002B/576

* 9 7 8 1 4 8 9 9 0 3 7 0 9 *